DEVELOPMENT OF SOFT PARTICLE CODE (SPARC)

T0135533

ADVANCES IN GEOTECHNICAL ENGINEERING AND
TUNNELLING

21

General editor:

D. Kolymbas

University of Innsbruck, Division of Geotechnical and Tunnel Engineering

In the same series (A.A.BALKEMA):

1. D. Kolymbas (2000), *Introduction to hypoplasticity*, 104 pages, ISBN 90-5809-306-9

2. W. Fellin (2000), *Rütteldruckverdichtung als plastodynamisches Problem (Deep vibration compaction as a plastodynamic problem)*, 344 pages, ISBN 90-5809-315-8

3. D. Kolymbas & W. Fellin (2000), *Compaction of soils, granulates and powders - International workshop on compaction of soils, granulates, powders*, Innsbruck, 28-29 February 2000, 344 pages, ISBN 90-5809-318-2

In the same series (LOGOS):

4. C. Bliem (2001), *3D Finite Element Berechnungen im Tunnelbau (3D finite element calculations in tunnelling)*, 220 pages, ISBN 3-89722-750-9

5. D. Kolymbas, ed. (2001), *Tunnelling Mechanics, Eurosummerschool, Innsbruck, 2001*, 403 pages, ISBN 3-89722-873-4

6. M. Fiedler (2001), *Nichtlineare Berechnung von Plattenfundamenten (Nonlinear Analysis of Mat Foundations)*, 163 pages, ISBN 3-8325-0031-6

7. W. Fellin (2003), *Geotechnik - Lernen mit Beispielen*, 230 pages, ISBN 3-8325- 0147-9

8. D. Kolymbas, ed. (2003), *Rational Tunnelling, Summerschool, Innsbruck 2003*, 428 pages, ISBN 3-8325-0350-1

9. D. Kolymbas, ed. (2004), *Fractals in Geotechnical Engineering, Exploratory Workshop, Innsbruck, 2003*, 174 pages, ISBN 3-8325-0583-0

10. P. Tanseng (2006), *Implementation of Hypoplasticity for Fast Lagrangian Simulations*, 125 pages, ISBN 3-8325-1073-7.

11. A. Laudahn (2006), *An Approach to 1g Modelling in Geotechnical Engineering with Soiltron*, 197 pages, ISBN 3-8325-1072-9.

12. L. Prinz von Baden (2005), *Alpine Bauweisen und Gefahrenmanagement*, 212 pages, ISBN 3-8325-0935-6.

13. D. Kolymbas, A. Laudahn, eds. (2005), *Rational Tunnelling, 2nd Summerschool, Innsbruck 2005*, 291 pages, ISBN 3-8325-1012-5.

14. T. Weifner (2006), *Review and Extensions of Hypoplastic Equations*, 240 pages, ISBN 978-3-8325-1404-4.

15. M. Mähr (2006), *Ground Movements Induced by Shield Tunnelling in Non-cohesive Soils*, 168 pages, ISBN 978-3-8325-1361-0.

16. A. Kirsch (2009), *On the Face Stability of Shallow Tunnels in Sand*, 178 pages, ISBN 978-3-8325-2149-3.

17. D. Renk (2011), *Zur Statik der Bodenbewehrung*, 165 pages, ISBN 978-3-8325-2947-5.

18. B. Schneider-Muntau (2013), *Zur Modellierung von Kriechhängen*, 225 pages, ISBN 978-3-8325-3474-5.

19. A. Blioumi (2014), *On Linear-Elastic, Cross-Anisotropic Rock*, 215 pages, ISBN 978-3-8325-3584-1.

20. G. Medicus (2015), *Barodesy and its Application for Clay*, 132 pages, ISBN 978-3-8325-4055-5.

Development of Soft Particle Code (SPARC)

Chien-Hsun Chen
University of Innsbruck, Division of Geotechnical and Tunnel Engineering

E-mail: Chien-Hsun.Chen@uibk.ac.at
Homepage: http://www.uibk.ac.at/geotechnik

The first three volumes have been published by Balkema
and can be ordered from:

A.A. Balkema Publishers
P.O.Box 1675
NL-3000 BR Rotterdam
e-mail: orders@swets.nl
website: www.balkema.nl

This publication has been produced with financial support of
the Vice Rector for Research of the University of Innsbruck.

Bibliographic information published by the Deutsche Nationalbibliothek

The Deutsche Nationalbibliothek lists this publication in the Deutsche
Nationalbibliografie; detailed bibliographic data are available in the
Internet at http://dnb.d-nb.de.

ISBN 978-3-8325-4070-8
ISSN 1566-6182

Logos Verlag Berlin GmbH
Comeniushof, Gubener Str. 47,
10243 Berlin
Tel.: +49 030 42 85 10 90
Fax: +49 030 42 85 10 92
INTERNET: http://www.logos-verlag.de

Acknowledgment

This work would not have been possible without the support of Prof. Dimitrios Kolymbas. He has been inspiring me with his unique scientific insight to all problems we encountered. I sincerely thank him for his guidance, patience, encouragement, as well as his unlimited office hours during my learning process as a doctoral student. Besides, I appreciate very much his good sense of humor and being considerate and supportive in many occasions, as a great teacher, mentor and friend.

I would also like to thank my second supervisor Prof. Wolfgang Fellin for his constructive suggestions and many inspirational discussions. His advice has largely improved the integrity of this work.

I sincerely thank Prof. Manuel Pastor for his participation in the committee of my doctoral dissertation defense.

I am grateful to the Austrian Science Fund (FWF) for the financial support on transportation costs for the meetings in Kaiserslautern with Dr. Jörg Kuhnert and Dr. Isabel Ostermann (Fraunhofer Institute for Industrial Mathematics, ITWM) and with the team of Prof. Vrettos (Technical University of Kaiserslautern) during the project I 547-N13. The discussions in the meetings with the groups were very constructive and helpful.

Thanks to *Vice Rector for Research of the University of Innsbruck* for their financial support in printing costs of this work.

It has been a great pleasure to work at the Division of Geotechnical and Tunnel Engineering. For me, this place has been not just a working place through these years, but also like a family, with smiles, laugh and support. Particularly, I should thank Anastasia Blioumi, Ansgar Kirsch, Barbara Schneider-Muntau, Daniel Renk, Gertraud Medicus, Iliana Polymerou, and Iman Bathaeian (alphabetically) for the collaboration and help with all tasks at work, as well as helpful, inspiring, and interesting discussions. I also owe many thanks to our secretary Sarah-Jane Loretz-Theodorine and to Mrs. Neuwirt for their kind and considerate assistance; to Stefan Tilg, Franz Berger and Franz Haas for helping with laboratory experiments. My life in Tyrol would not have been so delightful without their company.

To all my friends, here in Austria, back home in Taiwan, and far away in the States: You might not have noticed, but your friendship has enriched my life. Thank you!

To my family, for their love, support and understanding that I could not always be with them. My heartfelt thanks.

Kurzfassung

Das Ziel dieser Arbeit ist, eine numerische Simulationsmethode, den Soft PARticle Code (SPARC), zu entwickeln. Der Begriff "soft" hebt hervor, dass keine festen Partikel betrachtet werden (wie bei DEM). Die Unbekannten im Programm SPARC sind die Geschwindigkeiten der Partikel. Polynom-Interpolationen werden verwendet, um die räumlichen Ableitungen (z.B. des Stretchingtensors und der Divergenz des Cauchy'schen Spannungentensors) unter Berücksichtigung der Information der angrenzenden Partikeln, zu bilden. Die Barodesie, ein Materialmodell für granulare Stoffe in Ratenformulierung, wird zur Beschreibung der Spannungs-Verzerrungsbeziehung verwendet. Das nichtlineare Gleichungssystem, bestehend aus den Gleichgewichtsbedingungen, wird in SPARC iterativ gelöst und kann parallelisiert werden. Durch die Simulationen von Laborversuchen (Ödometerversuche, Biaxialversuche und Triaxialversuche) werden die Anwendungen und die Einschränkungen der aktuellen SPARC Version gezeigt. Zusätzlich zu den Simulationen von Standard Laborversuchen wurden Modellversuche mit Feinsand durchgeführt: Eine zyklische Kippbewegung einer Stützwand erzeugt eine bestimmte Bewegung in der Hinterfüllung (Sand). Die Sandpartikel bewegen sich entlang geschlossener Trajektorien. Die Bewegung wurde optisch aufgezeichnet und mit der "Particle Image Velocimetry" augewertet.

Abstract

This work aims at developing a numerical simulation method, Soft PARticle Code (SPARC). The term *soft* emphasizes that no boundaries between particles are defined and every particle possesses a support consisting of a set of adjacent particles. The unknowns in SPARC are velocities of particles. The polynomial interpolation/approximation method is utilized for the evaluation of spatial derivatives (such as stretching tensors and the divergence of Cauchy stress tensors) using the information carried by particles in supports. A novel nonlinear constitutive model, Barodesy, is adopted to describe the stress-strain relationship (of rate type) of granular materials. The system of equations consisting of spatial derivatives is solved using an iterative nonlinear solver and the computation of the Jacobian matrix is parallelized. The simulations of laboratory oedometer tests, biaxial tests, triaxial tests, and laboratory (zig-zag) model tests have been carried out to show the applications and limitations of the current version of SPARC. In addition to the simulations, laboratory (zig-zag) model tests using fine sand were carried out, in which the cyclic tilt of a retaining wall induces a peculiar motion in the backfill (sand), with closed trajectories (eddies). The motion of the backfill was optically traced and analyzed by means of particle image velocimetry.

Contents

Chapter 1

Introduction

1.1 Numerical simulation and approximation methods

Numerical simulations help to gain insight into complicated phenomena which can not be observed or measured. They also help to make predictions and, hence, are profitable in design. In numerical simulations, a continuum problem is discretized in space and time. Then, the governing equations (which are differential equations) are formulated and solved numerically. Numerical methods approximate the derivatives occurring in the differential equations based on numerical interpolation or approximation methods. In this way, the differential equations can be solved subject to some initial and/or boundary conditions.

Up to now, two numerical methods have primarily been used in civil engineering and other fields: the finite difference (FD) and the finite element (FE) methods. Over the past decades, they have been widely developed and adopted in commercial software, e.g., ABAQUS[1], PLAXIS, and FLAC. PLAXIS and FLAC are well-known in the field of geotechnical engineering. FE methods have been successfully applied to simulate problems with relatively small strains. Furthermore, the boundaries can be presented accurately. The drawback of the mesh-based FE methods is that the connectivities of the mesh nodes do not change during simulation. This results in considerable mesh distortion in Lagrangian[2] type formulations [36].

Meshless or meshfree approximation methods[3] has attracted great attention since the 1990s. In contrast to the traditional approaches, in meshless methods, the simulation domain is discretized by *particles* carrying physical quantities. The connectivities between particles are determined by an influence area, also referred to as support of a particle.[4] All particles in the domain of influence are termed *neighbors*. The

---

[1]Due to the popularity of the smoothed particle hydrodynamics methods, it has been implemented in ABAQUS in the latest version.

[2]FE methods using a deforming mesh to present the deforming continuum fall into this category. Meshless methods also belong to this category. In Eulerian methods, the governing equations are defined at fixed points in space (e.g. finite volume method).

[3]Both names are often seen in the literature, e.g., *meshless methods* in [3] and *meshfree methods* in [19].

[4]Also termed the *domain of influence* [3].

spatial derivatives of physical quantities in the domain of influence are evaluated by means of approximation techniques.[5] Therefore, these methods require no mesh at all. As particles move, their neighbors may vary. Hence, large deformations are easily handled within this framework. Among meshfree methods, the Smoothed Particle Hydrodynamics (SPH) introduced by Gingold, Lucy, and Monaghan [21, 40] for astrophysical problems is one of the earliest. In the past two decades, a large variety of meshless methods have been developed, such as the diffuse element method (DEM)[6], the reproducing kernel particle method (RKPM), the hp-cloud method, the partition of unity method (PUM), the meshless local Petrov-Galerkin method (MLPG), radial basis function methods (RBF), etc. These methods are different in the underlying approximation techniques, numerical solutions schemes for the governing equations (i.e. weak or strong from), time integration schemes, etc. Detailed summaries of recent developments and applications of meshless methods can be found in [3, 19, 25, 36, 37, 38]. As SPH methods have been widely applied in computational mechanics, HOOVER and his colleagues [25, 34, 49] called these methods *Smooth Particle Applied Mechanics* (SPAM). In SPAM, SPH is utilized for approximations using a normalized weight function and Newtonian mechanics governs the particle motion. The explicit fourth-order Runge-Kutta time integration is adopted for better numerical stability. It has been successfully applied to nonlinear dynamic problems such as molecular dynamics, solid mechanics (plastic deformation), heat conduction in 1D, Rayleigh-Bénard flow[7] in 2D, etc. Recently, the group of PASTOR [5, 6] used the Taylor-Galerkin scheme[8] in SPH to model the behavior of geomaterials, taking into account the effects of pore pressure under dynamic conditions.

Another group of researchers have tried to take advantage of both Lagrangian and Eulerian methods. For example, the *marker and cell* method pioneered by HARLOW in the 1960s [22, 23] is used to model viscous incompressible fluids with a free surface. In the marker and cell method, a background mesh is employed to evaluate the spatial derivatives in the governing equations by means of an FD approximation. First, the physical quantities (pressure and velocity in their case) are mapped to the background mesh. Secondly, the variables are determined by solving

---

[5][15] provides a good reference to approximations using moving least squares, radial basis functions, partition of unity methods, etc.

[6]Originally, it was proposed by Nayroles et al. [47] and aims at replacing the local interpolations in FE methods by the moving least squares approximation. In 1994, BELYTSCHKO and his colleagues [2, 39] developed the element-free Galerkin method (EFG) as a modified version. It is also based on moving least squares approximations, but it adopts a background mesh to perform the numerical integration. Hence, it is not truly meshfree.

[7]Rayleigh-Bénard flow is a type of convection of a fluid.

[8]Usually, continuous problems are discretized in time and space. *Taylor-Galerkin* denotes that a Taylor expansion is used for the time discretization employing the Runge-Kutta time integration scheme. The Galerkin method is adopted for the formulation of the governing equations over the simulation domain. Their early work has shown that the Taylor-Galerkin SPH is suitable for dynamic analyses and also to capture strain localization [41, 42, 43, 44].

the governing equations (i.e. the NAVIER-STOKEs equations). Finally, time integration is performed on the markers. No permanent data are stored on the mesh. Hence, this framework is a Lagrangian-Eulerian method, as the governing equations are formulated on fixed points and then the solutions are mapped to the markers. Later, this method was further developed under different names, among them particle-in-cell (PIC) method ([8]) and material point method (MPM) [52]. COETZEE [13] showed that the PIC method is capable of modeling granular flow. Lately, Beuth [4] used a version of MPM categorized in the *arbitrary Lagrangian-Eulerian* method, in which the governing equations are solved on a *deforming* mesh. The boundary conditions are applied through the FE method by using interface elements.[9] STOCK [51] also adopted the Galerkin PIC method with a fourth-order explicit RUNGE-KUTTA scheme for time integration and successfully simulated zonal flow in an experimental fusion-reactor.

1.2 Goal and outline

This work aims at developing a method based on the approximation of spatial derivatives by polynomial interpolation/approximation: A continuum is firstly discretized into 'soft' particles, which are mass points. The discretization turns a problem with infinite degrees of freedom into a problem with finite degrees of freedom that can be handled. The term *soft* emphasizes that no specified boundaries between particles are defined and every particle possesses a support consisting of a set of adjacent particles. The unknowns are the velocity vectors of the particles. The spatial derivatives (of the velocity field and the stress field) that appear in the governing equations and/or equations for kinematic boundary conditions are estimated for all particles by polynomial interpolation/approximation using the information stored in their supports. The system of equations is then solved using an iterative nonlinear solver. Accordingly, this method uses strong formulation and neither applies Galerkin nor seeks for weak solutions (i.e., no spatial integration is needed). As the governing equations are evaluated at each soft particle, it can be considered as a collocation method. The code of this framework is named *Soft Particle Code*, abbreviated as SPARC.

The scope of this work is listed in the following:

- Development of a method adopting the polynomial interpolation/approximation for the evaluation of spatial derivatives.
- Development of the soft particle code from the scratch.

---

[9]Unlike the volume elements that couple with material points, the interface element does not use material points.

- Determination of strong solutions by solving systems of equations using an iterative numerical solver in each time increment. The forward EULER method is adopted as the time advance scheme.
- Parallelization of the algorithm in the code to take advantage of parallel computing.
- Adoption of an advanced constitutive model, barodesy, for simulations of granular materials (soil).
- Test and documentation of the limitations of the code.

- Conduction and documentation of an experiment (grain motion subject to cyclic tilts of a wall hinged on the bottom).
- Simulation of the experimental test with SPARC.

The outline of this work is as follows. Chapter 2 is devoted to the framework of the proposed method, including the formulation of the governing equations, the evaluation of spatial derivatives, iterative solvers, kinematic boundary conditions at particles, etc. In Chapter 3, the simulations of the biaxial test are used to investigate the influences of (i) the configuration of the soft particles (regular and irregular arrays of particles), (ii) the neighbor search methods (fixed-radius search method and k-nearest neighbor search), (iii) the size of the support and the order of polynomials used for the evaluation of spatial derivatives, and (iv) the performance of the implemented numerical solvers. In Chapter 4, the simulations of two conventional laboratory tests, oedometer and triaxial CD tests, are detailed. The validation of the simulation results is carried out and documented in Chapter 5. To examine the ability of modeling large deformations, a simulation of a laboratory model test, in which large deformation is involved, is carried out and the results are detailed in Chapter 6. This experiment considers fine sand subject to cyclic tilts of a retaining wall hinged on the bottom. The grains exhibit closed trajectories around the active slip surface. The experimental set-ups, and the interpretation of the results of the model test are presented in Appendix J. Limitations of SPARC and recommendations for future developments are given in Chapter 7.

Chapter 2

Soft Particle Code (SPARC)

2.1 Introduction

The Soft PARticle Code (SPARC) is a particle-based numerical simulation method. In SPARC, a problem is discretized into time and space. A continuum is represented by a finite number of mass points carrying physical information, such as density (ρ), void ratio (e), Cauchy stress tensor, velocity ($\mathbf{v}$), position ($\mathbf{x}$), etc.[1] The word *soft* indicates that the boundaries between particles are not conceived as those in discrete element methods, in which every particle has an exact size, shape and boundary, and contact laws govern interactions between particles in contact. Instead, the motion of every particle is governed by the Cauchy's equation of motion:

$$\nabla \cdot \mathbf{T} + \rho \mathbf{g} = \mathbf{b} \tag{2.1}$$

where $\mathbf{g}$ is the mass force (usually: gravity) and $\mathbf{b}$ is the acceleration. For quasi-static problems, the equilibrium equation reads:

$$\nabla \cdot \mathbf{T} + \rho \mathbf{g} = \mathbf{0} \tag{2.2}$$

An alternative to eq. (2.2) is to take the time derivative of it:

$$\nabla \cdot \dot{\mathbf{T}} + \dot{\rho} \mathbf{g} = \mathbf{0} \tag{2.3}$$

Discretization of a continuum using soft particles allows the formulation of governing equations. Consider n_{p} particles in space with position vectors $\mathbf{x}_{(i)}$, $i = 1 \cdots n_{\mathrm{p}}$. A support of a particle i contains the particles near the particle i. It can be defined either by a radius r around $\mathbf{x}_{(i)}$ or by a fixed number (k) of the nearest particles to

---

[1] Parts of the descriptions in this section have been published in: I. Ostermann, J. Kuhnert, D. Kolymbas, C.-H. Chen, I. Polymerou, V. Šmilauer, C. Vrettos, D. Chen.(2013). "Meshfree generalized finite difference methods in soil mechanics—part I: theory", *Int J Geomath*, Springer-Verlag Berlin Heidelberg.

$\mathbf{x}_{(i)}$. Information carried in a support provides the basis for the evaluation of spatial derivatives using the polynomial interpolation/approximation method.[2]

In SPARC, the velocities of particles are the unknowns. They are determined by solving the above governing equations subject to boundary conditions using a numerical solver. This framework mainly consists of the two steps:

Step 1 - Evaluation of spatial derivatives:
Eq. (2.2), for example, is estimated numerically for every particle as follows: The velocity gradient $\mathbf{L}$ is firstly calculated from the interpolated velocity field $\mathbf{v}^t$ in the support:

$$\mathbf{L} = \nabla \mathbf{v}^t \tag{2.4}$$

The stretching tensor $\mathbf{D}$ and spin tensor $\mathbf{W}$, accounting for the deformation and rotation rates, are the symmetric and anti-symmetric parts of the velocity gradient:

$$\mathbf{D} = \frac{1}{2} \left(\mathbf{L} + \mathbf{L}^{\mathrm{T}} \right) \tag{2.5}$$

$$\mathbf{W} = \frac{1}{2} \left(\mathbf{L} - \mathbf{L}^{\mathrm{T}} \right) \tag{2.6}$$

The constitutive model, barodesy, accounting for the material behavior, is then used to predict the stress rate tensor $\mathring{\mathbf{T}}$. Adopting JAUMANN-ZAREMBA objective stress rate

$$\mathring{\mathbf{T}}^t = \dot{\mathbf{T}}^t - \mathbf{W}\mathbf{T}^t + \mathbf{T}^t \mathbf{W} \tag{2.7}$$

we obtain the stress rate for time integration

$$\dot{\mathbf{T}}^t = \mathring{\mathbf{T}}^t + \mathbf{W}\mathbf{T}^t - \mathbf{T}^t \mathbf{W} \tag{2.8}$$

and the stress tensor at time $t + \Delta t$:

$$\mathbf{T}^{t+\Delta t} = \mathbf{T}^t + \dot{\mathbf{T}}^t \Delta t \tag{2.9}$$

Step 2 - Solving the system of equations:
In eq. (2.2), the term $\nabla \cdot \mathbf{T}^{t+\Delta t}$ at every particle is calculated using the $\mathbf{T}^{t+\Delta t}$ field. Thereafter, a system of equations is constructed using eq. (2.2) for all particles. For problems in an n-dimensional space, there are n unknown

---

[2]When the reconstructed function from a set of point in space passes exactly through these points, it is called interpolation. When the reconstructed function does not pass through these points, but the sum of the squares of residuals is minimized, then it is referred to as an approximation.

velocity components at each particle. Eq. (2.2) provides also n equations at every particle. The equation system is solved using a numerical solver subject to boundary conditions.

This procedure shows that SPARC is a meshless collocation method using strong formulation.

The time integration scheme in SPARC is the forward EULER method. The parameters at time $t + \Delta t$ are computed after obtaining solutions of the unknown velocities $\mathbf{v}^t$ by solving the boundary value problem (in **Step 2** above). Thus, for particle i:

$$\mathbf{x}_{(i)}^{t+\Delta t} = \mathbf{x}_{(i)}^t + \mathbf{v}_{(i)}^t \Delta t \tag{2.10}$$

$$\begin{aligned}
e_{(i)}^{t+\Delta t} &= e_{(i)}^t + \dot{e}_{(i)}^t \Delta t \\
&= e_{(i)}^t + \left(1 + e_{(i)}^t\right) \left(\nabla_{(i)} \cdot \mathbf{v}^t\right) \Delta t \\
&= e_{(i)}^t + \left(1 + e_{(i)}^t\right) (\mathrm{tr}\mathbf{D})_{(i)} \Delta t
\end{aligned} \tag{2.11}$$

From the balance of mass, $\dot{\rho}_{(i)}^{t+\Delta t} + \rho_{(i)}^t \nabla \cdot \mathbf{v}^t = 0$ and $\nabla \cdot \mathbf{v}^t = \mathrm{tr}\mathbf{D}$, we have:

$$\rho_{(i)}^{t+\Delta t} = \rho_{(i)}^t - \rho_{(i)}^t (\mathrm{tr}\mathbf{D})_{(i)} \Delta t \tag{2.12}$$

For the time advance of stress tensor $\mathbf{T}^{t+\Delta t}$, follow eqs. (2.4) through (2.9).

The aforementioned framework of SPARC is illustrated in the flow chart in Fig. 2.1. Each step is detailed in the following sections of this chapter.

It is particularly noted the simulation is a *geometric-linear* analysis when the evaluation of the spatial derivative for the velocity gradient $\mathbf{L}$ is carried out in the *undeformed* particle configuration $\mathbf{x}^t$:

$$\mathbf{L}^t = \nabla^t \mathbf{v}^t \rightsquigarrow \mathbf{D}^t, \ \mathbf{W}^t \tag{2.13}$$

It is referred to as a *geometric-nonlinear* analysis when $\mathbf{L}$ is evaluated in the *deformed* configuration $\mathbf{x}^{t+\Delta t}$:

$$\mathbf{L}^{t+\Delta t} = \nabla^{t+\Delta t} \mathbf{v}^t \rightsquigarrow \mathbf{D}^{t+\Delta t}, \ \mathbf{W}^{t+\Delta t} \tag{2.14}$$

Due to the high non-linearity of the barodesy, using geometric-nonlinear analysis can cause unnecessary convergence problem. Therefore, the *geometric-linear* analysis is used in this work, unless stated otherwise.

Since the equilibrium condition must be fulfilled continuously, it is to evaluate

$$\nabla^{t+\Delta t} \cdot \mathbf{T}^{t+\Delta t} = \mathbf{0} \tag{2.15}$$

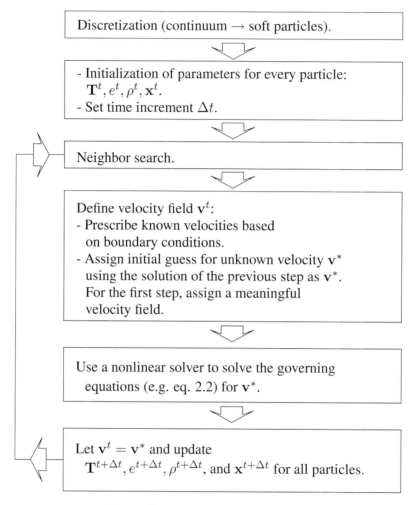

Figure 2.1: Flow chart of SPARC.

such that the equilibrium is satisfied at the particle configuration $\mathbf{x}^{t+\Delta t}$ as well.

2.2   Discretization of continuum

The discretization of a continuum allows a problem with infinite degrees of freedom to be reformulated into a problem with finite degrees of freedom. The continuum can be either discretized into a regular array of soft particles or into an irregular array of soft particles, i.e. a point cloud. The method for the generation of the point cloud used in this work is detailed in Appendix A. The influence of the configuration of particles is studied using the simulations of biaxial tests, as documented in Section 3.4.

2.3   Neighbor search

The properties of particle supports rely on the neighbor search method. Two neighbor search methods are studied herein: Fixed-radius neighbor search and k-nearest neighbor search (k-nn search). They have the following properties:

Fixed-radius neighbor search : Any particles j located within a distance r from particle i ($|\mathbf{x}_{(i)} - \mathbf{x}_{(j)}| \leq r$) are neighbors of particle i. The numbers of neighbors can vary from particle to particle and the particles on the boundaries have significantly fewer neighbors, as illustrated in Fig. 2.2a.

k-**nn search** : The nearest k particles to $\mathbf{x}_{(i)}$ are neighbors of particle i. The particles on the boundary, particularly, have neighbors which have larger distances to the particle under evaluation than the neighbors of the particles that are not on the boundaries, as illustrated in Fig. 2.2b. A fast k-nn search method is developed and documented in Appendix B.[3]

The influence of using various neighbor search methods and the support sizes are investigated in section 3.3.

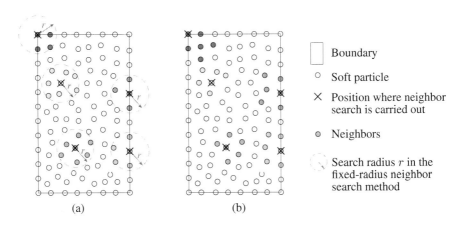

Figure 2.2: Examples of particle supports determined by (a) the fixed-radius search method and (b) the k-nn search method with 6 neighbors. The same particle configuration is used in both figures.

---

[3]Compared to the computational efforts for computing the stiffness matrix in the Newton's method, the computational costs for neighbor search are negligible.

2.4   Spatial derivatives

The evaluation of the spatial derivatives of the velocity ($\mathbf{v}$) field and the stress ($\mathbf{T}$) field is required to form the governing equation. It is demonstrated in this section how

$$\nabla \mathbf{v} = \frac{\partial v_q}{\partial x_s} = v_{q,s} = \begin{bmatrix} \frac{\partial v_1}{\partial x_1} & \frac{\partial v_1}{\partial x_2} & \frac{\partial v_1}{\partial x_3} \\ \frac{\partial v_2}{\partial x_1} & \frac{\partial v_2}{\partial x_2} & \frac{\partial v_2}{\partial x_3} \\ \frac{\partial v_3}{\partial x_1} & \frac{\partial v_3}{\partial x_2} & \frac{\partial v_3}{\partial x_3} \end{bmatrix} \quad q = 1 \cdots 3, s = 1 \cdots 3 \qquad (2.16)$$

and

$$\nabla \cdot \mathbf{T} = \frac{\partial T_{sq}}{\partial x_s} = T_{sq,s} = \begin{bmatrix} \frac{\partial T_{11}}{\partial x_1} + \frac{\partial T_{21}}{\partial x_2} + \frac{\partial T_{31}}{\partial x_3} \\ \frac{\partial T_{12}}{\partial x_1} + \frac{\partial T_{22}}{\partial x_2} + \frac{\partial T_{32}}{\partial x_3} \\ \frac{\partial T_{13}}{\partial x_1} + \frac{\partial T_{23}}{\partial x_2} + \frac{\partial T_{33}}{\partial x_3} \end{bmatrix} \quad q = 1 \cdots 3, s = 1 \cdots 3 \quad (2.17)$$

are computed at $\mathbf{x}_{(i)}$ (the position of particle i, $i = 1 \cdots n_\mathrm{p}$) using their supports and Cartesian coordinates x_1, x_2, x_3.

2.4.1   Polynomial interpolation/approximation

In SPARC, the spatial variation of the parameter of interest, f, in a support at $\mathbf{x}_{(i)}$ consisting of n_n neighbors, is evaluated by polynomial interpolation/approximation. Namely, f is approximated by a polynomial $\hat{f}$. Thereafter, the spatial derivatives of f can be evaluated. Herein, first order polynomials in a 2D space

$$\hat{f} = a_1 x_1 + a_2 x_2 \qquad (2.18)$$

are used as an example. Polynomials of higher order in a n-D space can be introduced analog to this example.

Say, there are totally n_p particles in a problem, as illustrated in Fig. 2.3. Given position vectors $\mathbf{x}_{(i)} = \begin{bmatrix} x_{1(i)} & x_{2(i)} \end{bmatrix}$ ($i = 1 \cdots n_\mathrm{p}$) and the parameters of interest $f_{(i)}$ of n_p particles, the spatial derivatives are calculated as follows. Consider *one* particle, particle i, the data used for the calculation are firstly collected from its neighbors:

$$\mathbf{F}_{(i)} = \begin{bmatrix} f_{(1)} \\ f_{(2)} \\ \vdots \\ f_{(n_\mathrm{n})} \end{bmatrix} \quad \text{and} \quad \boldsymbol{\mathcal{X}}_{(i)} = \begin{bmatrix} x_{1(1)} & x_{2(1)} \\ x_{1(2)} & x_{2(2)} \\ \vdots & \vdots \\ x_{1(n_\mathrm{n})} & x_{2(n_\mathrm{n})} \end{bmatrix} \qquad (2.19)$$

where the integer numbers $1, 2, \cdots, n_\mathrm{n}$ in the brackets $(\cdot)$ denote the *neighbors* of $\mathbf{x}_{(i)}$. Note that particle i is also in the support.

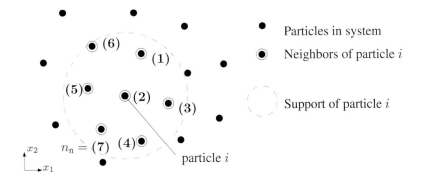

Figure 2.3: Illustration of neighbors of particle i. The numbers in $(\cdot)$ are the numeration for neighbors and $n_\mathrm{n} = 7$ is the total number of neighbors at particle i.

Second, to determine the coefficients a_1 and a_2 using the support of particle i, the origin is shifted:

$$
\bar{\mathbf{F}}_{(i)} = \begin{bmatrix} \left(f_{(1)} - f_{(i)}\right) \\ \left(f_{(2)} - f_{(i)}\right) \\ \vdots \\ \left(f_{(n_\mathrm{n})} - f_{(i)}\right) \end{bmatrix} = \underbrace{\begin{bmatrix} f_{(1)} \\ f_{(2)} \\ \vdots \\ f_{(n_\mathrm{n})} \end{bmatrix}}_{\mathbf{F}_{(i)}} - \begin{bmatrix} f_{(i)} \\ f_{(i)} \\ \vdots \\ f_{(i)} \end{bmatrix} \tag{2.20}
$$

and

$$
\bar{\boldsymbol{\mathcal{X}}}_{(i)} = \begin{bmatrix} \left(x_{1(1)} - x_{1(i)}\right) & \left(x_{2(1)} - x_{2(i)}\right) \\ \left(x_{1(2)} - x_{1(i)}\right) & \left(x_{2(2)} - x_{2(i)}\right) \\ \vdots & \vdots \\ \left(x_{1(n_\mathrm{n})} - x_{1(i)}\right) & \left(x_{2(n_\mathrm{n})} - x_{2(i)}\right) \end{bmatrix} = \underbrace{\begin{bmatrix} x_{1(1)} & x_{2(1)} \\ x_{1(2)} & x_{2(2)} \\ \vdots & \vdots \\ x_{1(n_\mathrm{n})} & x_{2(n_\mathrm{n})} \end{bmatrix}}_{\boldsymbol{\mathcal{X}}_{(i)}} - \begin{bmatrix} x_{1(i)} & x_{2(i)} \\ x_{1(i)} & x_{2(i)} \\ \vdots & \vdots \\ x_{1(i)} & x_{2(i)} \end{bmatrix}
$$

$$\tag{2.21}$$

With the relation $\bar{\mathbf{F}}_{(i)} = \bar{\boldsymbol{\mathcal{X}}}_{(i)}\mathbf{a}$ (eq. 2.18), the coefficients can be determined:

$$
\mathbf{a} = \bar{\boldsymbol{\mathcal{X}}}_{(i)}^{-1}\bar{\mathbf{F}}_{(i)} \tag{2.22}
$$

In the case that $\bar{\boldsymbol{\mathcal{X}}}_{(i)}$ is not a square matrix, the pseudo-inverse can be used for evaluating $\bar{\boldsymbol{\mathcal{X}}}_{(i)}^{-1}$. The coefficients a_1 and a_2 are the evaluated spatial derivatives of f

at $\mathbf{x}_{(i)}$:

$$\begin{bmatrix} \frac{\partial f}{\partial x_1} \\ \frac{\partial f}{\partial x_2} \end{bmatrix} \approx \begin{bmatrix} \frac{\partial \hat{f}}{\partial x_1} \\ \frac{\partial \hat{f}}{\partial x_2} \end{bmatrix} = \begin{bmatrix} a_1 \\ a_2 \end{bmatrix} \tag{2.23}$$

Note that this method forces $\hat{f}$ passing through particle i, as the constant term in the polynomial vanishes.

When $\hat{f}_{(k)} = f_{(k)}$ is valid for all neighbors in the supports ($k = 1 \ldots n_{\mathrm{n}}$), this method is referred to as *interpolation*. When $\hat{f}_{(k)} \neq f_{(k)}$, for $k = 1 \ldots n_{\mathrm{n}}$, but the sum of the squares of residuals $\sum_i^{n_{\mathrm{n}}} \left(\hat{f}_{(k)} - f_{(k)} \right)^2$ is minimized, it is referred to as *approximation*.

2.4.2 Evaluation of velocity gradient $\mathbf{L} = \nabla \mathbf{v}$ at one particle

Analog to section 2.4.1, given n_{p} particles in 3D space with velocity vectors $\mathbf{v}_{(i)} = \begin{bmatrix} v_{1(i)} & v_{2(i)} & v_{3(i)} \end{bmatrix}$ ($i = 1 \cdots n_{\mathrm{p}}$) and position vectors $\mathbf{x}_{(i)} = \begin{bmatrix} x_{1(i)} & x_{2(i)} & x_{3(i)} \end{bmatrix}$ ($i = 1 \cdots n_{\mathrm{p}}$), $\nabla \mathbf{v}$ at $\mathbf{x}_{(i)}$ is obtained as

$$\nabla \mathbf{v} = \begin{bmatrix} \frac{\partial v_1}{\partial x_1} & \frac{\partial v_1}{\partial x_2} & \frac{\partial v_1}{\partial x_3} \\ \frac{\partial v_2}{\partial x_1} & \frac{\partial v_2}{\partial x_2} & \frac{\partial v_2}{\partial x_3} \\ \frac{\partial v_3}{\partial x_1} & \frac{\partial v_3}{\partial x_2} & \frac{\partial v_3}{\partial x_3} \end{bmatrix} = \mathbf{a}^{\mathrm{T}} = \begin{bmatrix} a_{11} & a_{21} & a_{31} \\ a_{12} & a_{22} & a_{32} \\ a_{13} & a_{23} & a_{33} \end{bmatrix} \tag{2.24}$$

with $\mathbf{a} = \bar{\boldsymbol{\mathcal{X}}}_{(i)}^{-1} \bar{\boldsymbol{\mathcal{V}}}_{(i)}$ evaluated using the information at neighbors:

$$\bar{\boldsymbol{\mathcal{V}}}_{(i)} = \underbrace{\begin{bmatrix} v_{1(1)} & v_{2(1)} & v_{3(1)} \\ v_{1(2)} & v_{2(2)} & v_{3(2)} \\ \vdots & \vdots & \vdots \\ v_{1(n_{\mathrm{n}})} & v_{2(n_{\mathrm{n}})} & v_{3(n_{\mathrm{n}})} \end{bmatrix}}_{\boldsymbol{\mathcal{V}}_{(i)}} - \begin{bmatrix} v_{1(i)} & v_{2(i)} & v_{3(i)} \\ v_{1(i)} & v_{2(i)} & v_{3(i)} \\ \vdots & \vdots & \vdots \\ v_{1(i)} & v_{2(i)} & v_{3(i)} \end{bmatrix} \tag{2.25}$$

and

$$\bar{\boldsymbol{\mathcal{X}}}_{(i)} = \underbrace{\begin{bmatrix} x_{1(1)} & x_{2(1)} & x_{3(1)} \\ x_{1(2)} & x_{2(2)} & x_{3(2)} \\ \vdots & \vdots & \vdots \\ x_{1(n_{\mathrm{n}})} & x_{2(n_{\mathrm{n}})} & x_{3(n_{\mathrm{n}})} \end{bmatrix}}_{\boldsymbol{\mathcal{X}}_{(i)}} - \begin{bmatrix} x_{1(i)} & x_{2(i)} & x_{3(i)} \\ x_{1(i)} & x_{2(i)} & x_{3(i)} \\ \vdots & \vdots & \vdots \\ x_{1(i)} & x_{2(i)} & x_{3(i)} \end{bmatrix} \tag{2.26}$$

where $\boldsymbol{\mathcal{V}}_{(i)}$ and $\boldsymbol{\mathcal{X}}_{(i)}$ are matrices consisting of velocity and position vectors of *neighbors* at particle i, respectively.

2.4.3 Evaluation of $\nabla \cdot \mathbf{T}$ at one particle

Analog to section 2.4.1, given n_p particles in 3D space with effective CAUCHY stress tensors

$$\mathbf{T}_{(i)} = \begin{bmatrix} T_{11(i)} & T_{12(i)} & T_{13(i)} \\ T_{21(i)} & T_{22(i)} & T_{23(i)} \\ T_{31(i)} & T_{32(i)} & T_{33(i)} \end{bmatrix} \quad i = 1 \cdots n_\mathrm{p} \tag{2.27}$$

and position vectors $\mathbf{x}_{(i)} = \begin{bmatrix} x_{1(i)} & x_{2(i)} & x_{3(i)} \end{bmatrix}$ $(i = 1 \cdots n_\mathrm{p})$. As the stress tensor $\mathbf{T}$ is symmetric, let

$$\bar{\boldsymbol{\mathcal{T}}}_{(i)} = \underbrace{\begin{bmatrix} T_{11(1)} & T_{12(1)} & T_{13(1)} & T_{22(1)} & T_{23(1)} & T_{33(1)} \\ T_{11(2)} & T_{12(2)} & T_{13(2)} & T_{22(2)} & T_{23(2)} & T_{33(2)} \\ \vdots & \vdots & \vdots & \vdots & \vdots & \vdots \\ T_{11(n_\mathrm{n})} & T_{12(n_\mathrm{n})} & T_{13(n_\mathrm{n})} & T_{22(n_\mathrm{n})} & T_{23(n_\mathrm{n})} & T_{33(n_\mathrm{n})} \end{bmatrix}}_{\boldsymbol{\mathcal{T}}_{(i)}} -$$

$$\begin{bmatrix} T_{11(i)} & T_{12(i)} & T_{13(i)} & T_{22(i)} & T_{23(i)} & T_{33(i)} \\ T_{11(i)} & T_{12(i)} & T_{13(i)} & T_{22(i)} & T_{23(i)} & T_{33(i)} \\ \vdots & \vdots & \vdots & \vdots & \vdots & \vdots \\ T_{11(i)} & T_{12(i)} & T_{13(i)} & T_{22(i)} & T_{23(i)} & T_{33(i)} \end{bmatrix} \tag{2.28}$$

where $\boldsymbol{\mathcal{T}}_{(i)}$ is a matrix consisting of stress components of *neighbors*; $T_{sq(i)}$ is a component of $\mathbf{T}_{(i)}$.

With $\mathbf{b} = \bar{\boldsymbol{\mathcal{X}}}^{-1} \bar{\boldsymbol{\mathcal{T}}}$ ($\bar{\boldsymbol{\mathcal{X}}}$ is computed using eq. 2.26), $\nabla \cdot \mathbf{T}$ at $\mathbf{x}_{(i)}$ is obtained as:

$$\nabla \cdot \mathbf{T} = \begin{bmatrix} \frac{\partial T_{11}}{\partial x_1} + \frac{\partial T_{12}}{\partial x_2} + \frac{\partial T_{13}}{\partial x_3} \\ \frac{\partial T_{12}}{\partial x_1} + \frac{\partial T_{22}}{\partial x_2} + \frac{\partial T_{23}}{\partial x_3} \\ \frac{\partial T_{13}}{\partial x_1} + \frac{\partial T_{23}}{\partial x_2} + \frac{\partial T_{33}}{\partial x_3} \end{bmatrix} = \begin{bmatrix} b_{11} + b_{22} + b_{33} \\ b_{12} + b_{24} + b_{35} \\ b_{13} + b_{25} + b_{36} \end{bmatrix} \tag{2.29}$$

2.5 Constitutive model

A constitutive model consists of equations that describe the stress-strain relation of a material. It is also known as constitutive equation. Among many of them, the novel constitutive model, barodesy for sand, introduced by Kolymbas [31, 30, 32], is adopted herein due to its simplicity and capability of taking into account barotropy and pyknotropy[4], as well as the stress hardening and softening phenomena.

With the function

$$\mathbf{R}(\mathbf{D}) = (\operatorname{tr}\mathbf{D}^0)\mathbf{1} + c_1 \exp(c_2\mathbf{D}^0) \tag{2.30}$$

where $\mathbf{D}$ is the stretching tensor, i.e. the symmetric part of the gradient of velocity field (eq. 2.4), $\mathbf{1}$ is an identity matrix, $\mathbf{D}^0 = \mathbf{D}/|\mathbf{D}| = \mathbf{D}/\sqrt{\operatorname{tr}(\mathbf{D}^\mathsf{T}\mathbf{D})}$; c_1, c_2 are model coefficients and the *exp* is a matrix exponential; Kolymbas [28] proposed a constitutive equation with the following form:

$$\mathring{\mathbf{T}} = h(\sigma) \cdot (f\mathbf{R}^0 + g\mathbf{T}^0) \cdot |\mathbf{D}| \tag{2.31}$$

where $\mathring{\mathbf{T}}$ is the effective rate of Cauchy stress tensor; with

$$h(\sigma) = |\mathbf{T}|^{c_3} \tag{2.32}$$
$$f = c_4 \operatorname{tr}(\mathbf{D}^0) + c_5(e - e_c) + c_6 \tag{2.33}$$
$$g = -c_6 \tag{2.34}$$
$$e_c = (1 + e_{c0}) \exp\left(\frac{\sigma^{1-c_3}}{c_4(1 - c_3)}\right) - 1 \tag{2.35}$$

proposed in [29], where $\sigma := |\mathbf{T}|$, $e_{c0} = e_c(\sigma = 0)$ is the critical void ratio at zero stress, and c_i $(i = 3\ldots 6)$ are model coefficients.[5] Barodesy is, therefore, a function of the effective CAUCHY stress tensor $\mathbf{T}$ (unit in kPa), stretching tensor $\mathbf{D}$ and void ratio e:

$$\mathring{\mathbf{T}} = \mathcal{B}(\mathbf{T}, \mathbf{D}, e) \tag{2.36}$$

The coefficients of the model used in this paper, calibrated for Hostun sand, are listed in Table 2.1. For calibration of the model, readers are referred to [31, 30].

Table 2.1: Coefficients in barodesy for Hostun sand [29].

c_1	c_2	c_3	c_4	c_5	c_6	e_{c0}
-1.7637	-1.0249	0.5517	-1174	-4175	2218	0.8703

---

[4]Both stress level and density affect the mechanical behavior of soils. The influences of stress level and density are called *barotropy* and *pyknotropy*, respectively.

[5]c_3 is dimensionless and h has the unit kPa^{c_3}. c_4, c_5 and c_6 have the unit kPa$^{1-c_3}$. As a result, $\mathring{\mathbf{T}}$ has the unit kPa^{c_3} (kPa$^{1-c_3}$) = kPa.

2.6   System of equations

Given $\mathbf{T}^t$, e^t, ρ^t, $\mathbf{x}^t$, and $\mathbf{v}^t$ of all particles, the system of equations is obtained with the following three steps, and detailed in the next three subsections.

1. Evaluation of $\mathbf{T}^{t+\Delta t}$ at all particles.

2. Evaluation of $\nabla^{t+\Delta t} \cdot \mathbf{T}^{t+\Delta t}$ at all particles

3. Obtaining the system of equations

2.6.1   Evaluation of $\mathbf{T}^{t+\Delta t}$ at all particles

We firstly consider *one* particle in 3D space. Given velocity vectors[6] $\mathbf{v}^t$, position vectors $\mathbf{x}^t$, densities ρ, void ratios e, and stress states $\mathbf{T}^t$ of *all* particles in a system, the calculation of the stress state $\nabla \cdot \mathbf{T}^{t+\Delta t}$ at *one* particle follows the procedure:

⤳   Finding which particles are neighbors. (Section 2.3)

⤳   Evaluation of $\mathbf{L}^t = \nabla^t \mathbf{v}^t$ (Section 2.4.2)

⤳   Evaluation of stretching and spin tensors $\mathbf{D}^t$ and $\mathbf{W}^t$, respectively (eqs. 2.5, 2.6, and 2.13)

⤳   Evaluation of stress increment $\overset{\circ}{\mathbf{T}}{}^t$ by means of constitutive equations (Section 2.5)

⤳   Evaluation of objective stress rate tensor $\dot{\mathbf{T}}^t = \overset{\circ}{\mathbf{T}}{}^t - \mathbf{T}^t \mathbf{W}^t + \mathbf{W}^t \mathbf{T}^t$ (eq. 2.8)

⤳   Time advance of stress tensor $\mathbf{T}^{t+\Delta t} = \mathbf{T}^t + \dot{\mathbf{T}}^t \Delta t$. (eq. 2.9)

It can be seen that $\mathbf{D}^t$ and $\overset{\circ}{\mathbf{T}}{}^t$ at a particle consist of the following parameters:[7]

$$\mathbf{D}^t = \mathbf{D}^t \left(\mathbf{L}^t (\boldsymbol{\mathcal{X}}^t, \boldsymbol{\mathcal{V}}^t) \right) = \mathcal{D} \left(\boldsymbol{\mathcal{X}}^t, \boldsymbol{\mathcal{V}}^t \right) \tag{2.37}$$

$$\overset{\circ}{\mathbf{T}}{}^t = \mathcal{B} \left(\mathbf{T}^t, \mathbf{D}^t, e^t \right)$$

$$= \mathcal{B} \left(\mathbf{T}^t, \mathcal{D}(\boldsymbol{\mathcal{X}}^t, \boldsymbol{\mathcal{V}}^t), e^t \right) \tag{2.38}$$

$$= \mathcal{R} \left(\mathbf{T}^t, \boldsymbol{\mathcal{X}}^t, \boldsymbol{\mathcal{V}}^t, e^t \right) \tag{2.39}$$

where $\boldsymbol{\mathcal{X}}$ is a matrix consisting of position vectors of the particle's neighbors (see eq. 2.25) and $\boldsymbol{\mathcal{V}}$ consisting of velocity vectors of the particle's neighbors (see eq. 2.26). Consequently, $\mathbf{T}^{t+\Delta t}$ at a particle consists of the following parameters:

$$\mathbf{T}^{t+\Delta t} \left(\mathbf{T}^t, \boldsymbol{\mathcal{X}}^t, \boldsymbol{\mathcal{V}}^t, e^t, \Delta t \right) = \mathbf{T}^t + \dot{\mathbf{T}}^t \Delta t \tag{2.40}$$

---

[6]For the unknown velocity components, initial guesses can be assigned to them.

[7]Individual symbols are used herein to distinguish functions parameterized with different variables so that the mathematical descriptions are clear, as emphasized in [16].

Second, since the computation of $\nabla \cdot \mathbf{T}^{t+\Delta t}$ requires the field of $\mathbf{T}^{t+\Delta t}$, $\mathbf{T}^{t+\Delta t}$ at *all* particles must be evaluated prior to the evaluation of $\nabla \cdot \mathbf{T}^{t+\Delta t}$. Thus, the above procedure must be repeated to obtain $\mathbf{T}^{t+\Delta t}$ at all particles.

2.6.2   Evaluation of $\nabla^{t+\Delta t} \cdot \mathbf{T}^{t+\Delta t}$ at all particles

For the evaluation of the divergence of the stress tensor $\nabla^{t+\Delta t} \cdot \mathbf{T}^{t+\Delta t}$, position vectors $\mathbf{x}^{t+\Delta t}$ at *all* particles need to be advanced:

$$\mathbf{x}^{t+\Delta t}\left(\mathbf{x}^t, \mathbf{v}^t, \Delta t\right) = \mathbf{x}^t + \mathbf{v}^t \Delta t \tag{2.41}$$

Now, we consider only *one* particle. Given $\mathbf{T}^{t+\Delta t}$ and $\mathbf{x}^{t+\Delta t}$ at all particles, the vector $\nabla^{t+\Delta t} \cdot \mathbf{T}^{t+\Delta t}$ at this particle is evaluated using the information carried by its neighbors, as detailed in Section 2.4.3. Let $\mathcal{E}$ be the residual[8] of the vector of the equilibrium equation (eq. 2.2) at the particle, we then have

$$\mathcal{E}\left(\mathcal{T}^t, \mathcal{X}^t, \mathcal{V}^t, e^t, \Delta t, \rho^t, \boldsymbol{g}\right) = \nabla^{t+\Delta t} \cdot \mathbf{T}^{t+\Delta t} + \rho^t \boldsymbol{g} \tag{2.42}$$

where $\mathcal{T}$ is a matrix consisting of stress components of stress tensors $\mathbf{T}$ (see eq. 2.28) of neighbors and e is a vector consisting of void ratios of neighbors. $\mathcal{E}$ is a vector with three components ($\mathcal{E}_j = \partial T_{sj}^{t+\Delta t}/\partial x_s^{t+\Delta t} + \rho g_j, j = 1 \cdots 3, s = 1 \cdots 3$) which correspond to the three components of the velocity vector, v_j ($j = 1 \cdots 3$). If a component in v_j of a particle is unknown, the corresponding component in $\mathcal{E}_j$ at that particle needs to be calculated, and $\mathcal{E}_j = 0$ (for any j) must be fulfilled to obtain v_j.

Now, $\mathcal{E}$ must be evaluated for *all* n_p particles in a problem and $\mathcal{E}_{(i)}$ with $i = 1 \cdots n_\mathrm{p}$ are stored in one matrix, $\boldsymbol{\Xi}$:

$$\boldsymbol{\Xi} = \begin{bmatrix} \mathcal{E}_{(1)} \\ \mathcal{E}_{(2)} \\ \vdots \\ \mathcal{E}_{(n_\mathrm{p})} \end{bmatrix} = \begin{bmatrix} \mathcal{E}_{1(1)} & \mathcal{E}_{2(1)} & \mathcal{E}_{3(1)} \\ \mathcal{E}_{1(2)} & \mathcal{E}_{2(2)} & \mathcal{E}_{3(2)} \\ \vdots & \vdots & \vdots \\ \mathcal{E}_{1(n_\mathrm{p})} & \mathcal{E}_{2(n_\mathrm{p})} & \mathcal{E}_{3(n_\mathrm{p})} \end{bmatrix} \tag{2.43}$$

Similarly, for $i = 1 \cdots n_\mathrm{p}$, the velocity vectors $\mathbf{v}_{(i)}$, position vectors $\mathbf{x}_{(i)}$, stress tensors $\mathbf{T}_{(i)}$, void ratios $e_{(i)}$, and densities $\rho_{(i)}$ of *all* particles are stored in matrices

---

[8]When an iterative solver is used to solve a system of equations $\mathbf{f}(\mathbf{d}) = \mathbf{0}$, the left-hand side of the equation system, $\mathbf{f}(\mathbf{d})$, is called *residual* because during the iteration the values of residuals are non-zero and they change. When the residuals are close enough to zero, the equations are said to be solved for variables $\mathbf{d}$.

$\mathbb{V}$ ($n_p \times 3$ matrix), $\mathbb{X}$ ($n_p \times 3$ matrix), $\mathbb{T}$ ($n_p \times 6$ matrix[9]), $\mathbb{R}$ ($n_p \times 1$ matrix), and $\mathbb{E}$ ($n_p \times 1$ matrix), respectively. Ξ, therefore, contains the following parameters

$$\Xi = \Xi\left(\mathbb{T}^t, \mathbb{X}^t, \mathbb{V}^t, \mathbb{E}^t, \Delta t, \mathbb{R}^t, g\right) \tag{2.44}$$

It must be noted that $\mathbb{V}^t$ consists of both known and unknown velocities and there are totally d^{of} (degrees of freedom) unknown velocity components in $\mathbb{V}^t$.

Remark 1: To assist the explanation of the implementation of arc-length methods, it should be noted that , from eq. (2.41) we have:

$$\mathbf{v}^t\left(\mathbf{x}^t, \mathbf{x}^{t+\Delta t}, \Delta t\right) = \frac{\mathbf{x}^{t+\Delta t} - \mathbf{x}^t}{\Delta t} \tag{2.45}$$

This equation shows that $\mathbf{v}^t$ is parameterized with $\mathbf{x}^t$, $\mathbf{x}^{t+\Delta t}$, and Δt. As a result, eq. (2.44) can also be expressed as:

$$\begin{aligned} \mathbf{G} &= \mathscr{E}\left(\mathbb{T}^t, \mathbb{X}^t, \mathbb{V}^t\left(\mathbf{x}^t, \mathbf{x}^{t+\Delta t}, \Delta t\right), \mathbb{E}^t, \Delta t, \mathbb{R}^t, g\right) \\ &= \mathscr{G}\left(\mathbb{T}^t, \mathbb{X}^t, \mathbb{X}^{t+\Delta t}, \mathbb{E}^t, \Delta t, \mathbb{R}^t, g\right) \end{aligned} \tag{2.46}$$

in which unknowns are positions in $\mathbb{X}^{t+\Delta t}$ and there are d^{of} (degrees of freedom) unknown position components in $\mathbb{X}^{t+\Delta t}$. Eqs. (2.44) and (2.46) are residuals computed using *the same* equilibrium equation (eq. 2.2) . The former is parameterized using velocity (namely, the residuals are expressed using velocity as unknown; adopted in SPARC) and the latter is parameterized with positions (using positions as unknown; for the explanation of arc-length methods).

2.6.3 Obtaining the system of equations

The system of equations in SPARC reads:

$$\mathbf{y} = 0$$

The components in matrix Ξ (residuals) that correspond to the unknown components in matrix $\mathbb{V}$ consist in the vector of $\mathbf{y}$.

In order to obtain $\mathbf{y}$, two matrices are introduced:
(a) Matrix $\mathcal{D}^{of}$ consisting of integers 1 for *unknown* and 0 for *known*, with dimensions $n_p \times 3$, storing the information of the degrees of freedom of the system.[10]

---

[9]only 6 components of $\mathbf{T}$ are stored as $\mathbf{T}$ is symmetric

[10]The integers 1 and 0 may be replaced by the Boolean (logic) data type, with *true* for integer 1, and *false* for integer 0.

(b) Matrix $\mathcal{N}$ with dimensions $n_p \times 3$ stores the *matrix numeration* for matrices $\mathbb{V}$ and Ξ:

$$\mathcal{N} = \begin{bmatrix} 1 & n_p + 1 & 2n_p + 1 \\ 2 & n_p + 2 & 2n_p + 2 \\ \vdots & \vdots & \vdots \\ n_p & 2n_p & 3n_p \end{bmatrix} \tag{2.47}$$

The integer numbers $1, 2, \cdots, 3n_p$ in $\mathcal{N}$ tag the positions of components in a matrix so that a matrix can be transformed into a vector.

Given $\mathcal{D}^{of}$ and $\mathcal{N}$, the following algorithm is used to obtained a mapping vector $\mathbf{m}$.

```
counter = 0
for j = 1:3
    for i = 1:n_p
        If  𝒟^of(i,j) = 1
            counter = counter + 1;
            m(counter) = 𝒩(i,j)
        end
    end
end
```

$\mathbf{m}$, with dimensions $d^{of} \times 1$, stores the matrix numeration (defined in $\mathcal{N}$) corresponding to the component of $\mathcal{D}^{of}$ with a value of 1 (i.e. the corresponding velocity component in $\mathbb{V}$ is an unknown variable). Let i_e be the index for the numeration of the residuals $\mathbf{y} = y_{i_e}$ ($i_e = 1 \cdots d^{of}$), we obtain the numeration for the position (defined by $\mathcal{N}$) in Ξ:

$$y_{i_e} = \Xi\left(\mathbf{m}(i_e)\right) \tag{2.48}$$

An example of such operation is given in Appendix C. Thus, $\mathbf{m}$ maps the components between $\mathbf{y}$ (a vector) and Ξ (a matrix). This way, $\mathbf{y}$ is generated by *assembling* the residuals stored in Ξ. It consists of the following parameters (the same as Ξ, see eq. 2.44):

$$\mathbf{y} = \mathbf{y}\left(\mathbb{T}^t, \mathbb{X}^t, \mathbb{V}^t, \mathbb{E}^t, \Delta t, \mathbb{R}^t, \boldsymbol{g}\right) \tag{2.49}$$

With the same method, a vector ($\mathbf{u}$) that contains only unknown variables (v^t) of the system is obtained by *assembling* v^t from $\mathbb{V}^t$ via $\mathbf{m}$ (see Appendix C for an example):

$$\mathbf{u}^t = u_{i_u}^t = \mathbb{V}^t\left(\mathbf{m}(i_u)\right) \tag{2.50}$$

u is required in the Newton's method to compute the Jacobin matrix (see Section 2.7.1.2). Thus, the system of equations is obtained:

$$\mathbf{y}(\mathbf{u}^t) = \mathbf{0} \tag{2.51}$$

Eq. (2.51) is then solved using a numerical solver to determine $\mathbf{u}^t$.

Remark 2: This remark is added here to assist the comprehension of arc-length methods in a later section (Section 2.7.4). Recall that, in **Remark 1** (see Section 2.6.2), it has been shown that velocity can be parameterized with position. Analog to eq. (2.45),

$$\mathbf{u}^t = \frac{\mathbf{w}^{t+\Delta t} - \mathbf{w}^t}{\Delta t} \tag{2.52}$$

where $\mathbf{w}^{t+\Delta t}$ is a vector consisting of only unknown position components of $\mathbb{X}^{t+\Delta t}$, while $\mathbf{w}^t$ and Δt are given. Accordingly, the system of equations (eq. 2.51) may also be expressed using the unknown positions at time $t + \Delta t$.

$$\mathcal{Y}(\mathbf{w}^{t+\Delta t}) = \mathbf{0} \tag{2.53}$$

2.7   Numerical Solver

The Newton's method is an iterative solver often employed to solve a nonlinear system of equations. However, in the cases presented in this work, the Newton's method often failed to converge due to complex deformations (e.g. strain localization). Considerable efforts have been made to find a proper nonlinear solver for SPARC. Among many, the Marquardt's algorithm [45][11] and the pseudo-arc-length method are implemented in SPARC. Their capabilities of solving nonlinear systems in the problems involving strain localization phenomena are investigated and presented in Chapter 3. In addition, the computation of Jacobian matrix was limited to particles in the influence area so that the computing time is shortened. Moreover, in order to significantly improve the accuracy of the time integration for $\mathbf{T}^{t+\Delta t} = \mathbf{T}^t + \dot{\mathbf{T}}^t \Delta t$ without changing the overall simulation increment Δt, substeps and a fourth-order RUNGE-KUTTA method are implemented in the time integration of stress.

---

[11]It is recognized as the Levenberg-Marquardt method in the literature.

2.7.1 The Newton's method

2.7.1.1 One-variable problem

The procedure of solving the equation $y(u) = 0$ with one unknown variable u using the Newton's method[12] is illustrated in Fig. 2.4. The thick line illustrates the function

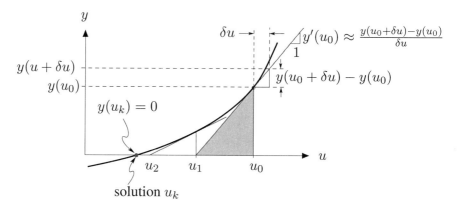

Figure 2.4: The Newton's method.

$y(u)$. Given the function $y(u)$, the Newton's method solves the equation $y(u) = 0$ iteratively as follows [33]:

1. An initial guess u_0 is selected for the solution and evaluate $y(u_0)$, where the subscript k ($k = 1, 2, \cdots$) is an iteration counter (Fig. 2.4).

2. The slope of the tangent line at u_0 is approximated by the quotient:

$$y'(u_o) \approx \frac{y(u_o + \delta u) - y(u_o)}{\delta u} \tag{2.54}$$

   where δu is a very small increment.[13]

3. Evaluation of u_1:

$$u_1 = u_o - y'(u_o)^{-1} y(u_o) \tag{2.55}$$

$$= u_o + \Delta u \tag{2.56}$$

   where $-y'(u_o)^{-1} y(u_o)$ is known as the *correction* [45] (denoted as Δu), to the iterates.

---

[12]The Newton's method is also known as the Newton-Raphson method [33].

[13]$\delta u = \sqrt{\text{eps}}$ is used herein, where 'eps' is the so-called machine epsilon which defines the precision of floating numbers. See the following for more information:
http://www.math.pitt.edu/~trenchea/math1070/MATH1070_2_Error_and_Computer_Arithmetic.pdf

4. Computation of $y(u_1)$.

5. Step 2 through step 4 are iterated for computing $y(u_2)$, $y(u_3)$, $\ldots y(u_k)$, $y(u_{k+1})$, until $y(u_{k+1}) \rightarrow 0$. In practice, iteration is stopped and an acceptable approximation to the exact solution is obtained when $|y(u_k)| \leq \epsilon$, where ϵ is a user-defined threshold, a small value characterizing the accuracy of the approximation.

This method can be extended to a system of equations with many variables.

2.7.1.2 Application in SPARC: Multi-variables

The system of equations (eq. 2.51) in SPARC is the collection of governing equations which must be solved for the unknown velocity components. It may also be written as:

$$y_{i_e}(\mathbf{u}^t) = 0 \qquad (2.57)$$

where $i_e = 1, \ldots, d^{of}$ (with 'e' denoting *equation*) and $\mathbf{u}^t = u^t_{i_u}$ ($i_u = 1, \ldots, d^{of}$, with 'u' denoting *unknown*).[14] See Appendix C for the method of how to obtain $\mathbf{u}^t$ by assembling unknown components in $\mathbb{V}$ using the mapping vector $\mathbf{m}$ (eq. 2.50, or see Appendix C for an example).

In the following eqs. (2.58) through (2.64), $u = u^t$ and $\mathbf{u} = \mathbf{u}^t$.

Similar to eq. (2.54), the influence of any component in u_{i_u} on y_{i_e}, in the k-th iteration, can be calculated:

$$
\begin{aligned}
[y_{i_e,i_u}]_k &= \left[\frac{\partial y_{i_e}(\mathbf{u})}{\partial u_{i_u}} \right]_k \\
&\approx \left[\frac{y_{i_e}(u_1, u_2, \ldots, u_{i_u} + \delta u, \ldots) - y_{i_e}(u_1, u_2, \ldots, u_{i_u}, \ldots)}{\delta u} \right]_k
\end{aligned}
\qquad (2.58)
$$

y_{i_e,i_u} is known as Jacobian matrix or stiffness matrix:

$$\mathbf{J} = y_{i_e,i_u} \qquad (2.59)$$

Applying eqs. (2.55) and (2.56) to this multi-variable case, $(u_{i_u})_{k+1}$ is obtained:

$$
\begin{aligned}
(u_{i_u})_{k+1} &= (u_{i_u})_k - [y_{i_e,i_u}]_k^{-1} \, y_{i_e}(\mathbf{u}_k) \qquad (2.60) \\
&= (u_{i_u})_k + \Delta \mathbf{u}_k \qquad (2.61)
\end{aligned}
$$

---

[14]In general, the number of governing equations does not necessarily equal the number of unknowns.

where $\Delta \mathbf{u}_k = - \left[y_{i_e, i_u} \right]_k^{-1} y_{i_e}(\mathbf{u}_k)$ is a vector of the corrections of the iterates.

The error of the approximated solutions can be defined by the Euclidean norm:

$$\delta_{k+1} = \sqrt{\sum_i^p \left(y_i(\mathbf{u}_{k+1}) \right)^2} \tag{2.62}$$

or alternatively, by the maximum error:

$$\delta_{k+1} = \max \left\{ |y_i(\mathbf{u}_{k+1})| \right\} \tag{2.63}$$

If

$$\delta_{k+1} \leq \epsilon \tag{2.64}$$

where ϵ is a pre-defined threshold, then the solution $\mathbf{u}_{k+1}$ is accepted. Otherwise, let $k \leftarrow k+1$ and iterate the calculations in equations (2.58) through (2.64). δ_{k+1} is known as the objective function value of the system that defines the wellness of the solution.

2.7.2   Reduced stiffness matrix

In SPARC, due to the usage of a support for evaluating the spatial derivatives for every particle, a variation of a velocity component does not affect the vector $\nabla^{t+\Delta t} \cdot \mathbf{T}^{t+\Delta t} + \rho \mathbf{g}$ at all particles. Thus, the calculation of zero values (in the Jabobian matrix) at places where it is known to have no influence is avoided. This idea is illustrated in Fig. 2.5. The vector $\nabla^{t+\Delta t} \cdot \mathbf{T}^{t+\Delta t} + \rho \mathbf{g}$ for particle 1 (Fig. 2.5a) is

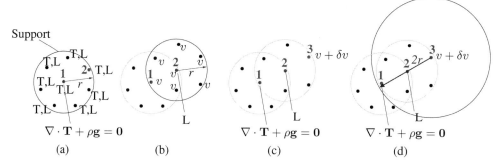

Figure 2.5: Illustration of the influenced area if a velocity component is changed. In the figures, $\mathbf{T}^{t+\Delta t}$ is referred to as $\mathbf{T}$, $\mathbf{L} = \mathbf{L}^t$.

evaluated using particles in its support (with radius r). In order to evaluate $\nabla^{t+\Delta t} \cdot \mathbf{T}^{t+\Delta t}$ at particle 1, $\mathbf{T}^{t+\Delta t}$ at neighbors of particle 1 must be priorly computed, e.g.

$\mathbf{T}^{t+\Delta t}$ at particle 2. To compute $\mathbf{T}^{t+\Delta t}$ at particle 2 (Fig. 2.5b), the velocity gradient $\mathbf{L}^t$ at particle 2 must be priorly computed and the velocity field in the support of particle 2 is used for that purpose. Consequently (see Fig. 2.5c), if a the velocity component v at, say, particle 3 is changed to $v + \delta v$, the vector $\nabla^{t+\Delta t} \cdot \mathbf{T}^{t+\Delta t} + \rho \mathbf{g}$ at the particles in the range of $2r$ of particle 3 is affected (Fig. 2.5d); and the stress tensor $\mathbf{T}^{t+\Delta t}$ at the particles in the range of $3r$ of particle 3 is affected. Therefore, while building the stiffness matrix $y_{i,j}$, considerable amount of computations can be avoided when only variations in velocity component v_j are calculated that affect equilibrium equations y_i. The so-obtained matrix is called reduced stiffness matrix.

The implementation of the above method is as follows. Recall that, subject to the influences of the variation of *one* component in $\mathbf{u}$ (see eq. 2.58), the residuals y_{i_e} ($i_e = 1 \cdots d^{of}$) (i.e. a column of the Jacobian matrix) are obtained using the computation procedure described in Section 2.6.1, 2.6.2, and 2.6.3. Before using the numerical solver, for every particle, three lists of neighbors obtained via fixed-radius neighbor search method using radius r, $2r$ and $3r$ are stored in three individual matrices. While computing a column of the Jacobian matrix, it is to firstly carry out the computations of $\mathbf{T}^{t+\Delta t}$ (described in Section 2.6.1) only for particles that belong to the neighbor list found with a radius of $3r$ away from the particle whose velocity component is changed to $u_{i_u} + \delta u$ (see eq. 2.58). Note that the mapping vector $\mathbf{m}$ correlates the vector $\mathbf{u}$ (assembling of unknown velocity components) and matrix $\mathbb{V}$ (velocities of all particles, including known and unknown velocity components). (See eq. 2.50.) Second, the computations of $\nabla^{t+\Delta t} \cdot \mathbf{T}^{t+\Delta t} + \rho \mathbf{g}$ (described in Section 2.6.2) are carried out only for particles that belong to the neighbor list found with a radius of $2r$ away from the particle whose velocity component is changed to $u_{i_u} + \delta u$ (see eq. 2.58). Finally, the column the Jacobian matrix is obtained by assembling the residuals from the Ξ matrix (eq. 2.43) using the mapping vector $\mathbf{m}$ (see eq. 2.48 or Appendix C).

Results: A three-dimensional oedometer simulation is used for testing the computing time of simulations with and without the above described technique. The results are listed in Table 2.2. The spatial derivatives are evaluated using one-dimensional first-order polynomials. The programs are compiled with Matlab Coder and one processor is used for computation. In the simulations in Group A, the reduced stiffness matrix method is used. In the simulations in Group B, all components in the stiffness matrix are computed. The results show that the computing times using reduced stiffness matrix are significantly shorter.

2.7.3 Marquardt's Algorithm

The Marquardt's algorithm [45] is implemented in SPARC to improve the convergence in the Newton's iteration. The algorithm is summarized in this section.

Table 2.2: Comparisons of the computation times in simulations between group A and B. The simulations in Group A compute the reduced stiffness matrix. The simulations in Group B compute all components in the stiffness matrix.

Group	Degrees of freedom	Number of neighbors	Computation time t	t_B/t_A
A	735	$4 \ldots 7$	116.76 sec	7.8
B	735	$4 \ldots 7$	913.14 sec	
A	5577	$4 \ldots 7$	3341.4 sec	18.57
B	5577	$4 \ldots 7$	62061 sec	

The Newton's method solves the system of equations $\mathbf{J}_k \Delta \mathbf{u}_k = -\mathbf{y}_k$ (see eqs. 2.59 through 2.61) in each iteration to obtain the correction of the solution set $\Delta \mathbf{u}_k$, where $\mathbf{J}$ is the Jacobian matrix $\partial y_i / \partial u_j$ and k is the iteration counter. The iterates are updated as:

$$\mathbf{u}_{k+1} = \mathbf{u}_k + \Delta \mathbf{u}_k \tag{2.65}$$

The corresponding Euclidean norm of the errors using the updated solution set $\mathbf{u}_{k+1}$ is $|\mathbf{y}_{k+1}|$. The solution is said to be found when $|\mathbf{y}_{k+1}|$ is smaller than an acceptable value ϵ. However, Newton's method does not ensure convergence, i.e. $|\mathbf{y}_{k+1}| \leq |\mathbf{y}_k|$. MARQUARDT [45] suggested the following algorithm that avoids divergence.

Since, in general, $\mathbf{J}$ is not a square matrix, the following operations in an iteration k of the Newton's method is suggested [45]:

$$\mathbf{J}_k \Delta \mathbf{u}_k = \quad -\mathbf{y}_k \tag{2.66}$$

$$\rightsquigarrow \quad \mathbf{J}_k^{\mathrm{T}} \mathbf{J}_k \Delta \mathbf{u}_k = -\mathbf{J}_k^{\mathrm{T}} \mathbf{y}_k \tag{2.67}$$

Let

$$\mathbf{A} = A_{ij} = \mathbf{J}_k^{\mathrm{T}} \mathbf{J}_k \tag{2.68}$$

$$\mathbf{g} = g_i = -\mathbf{J}_k^{\mathrm{T}} \mathbf{y}_k \tag{2.69}$$

and scale[15] $\mathbf{A}$ and $\mathbf{g}$ using the main diagonal of $\mathbf{A}$:

$$\mathbf{A}^* = A_{ij}^* \tag{2.70}$$

$$\mathbf{g}^* = g_i^* \tag{2.71}$$

---

[15]"This choice of scale causes the $\mathbf{A}$ matrix to be transformed into the matrix of simple correlation coefficients among $\frac{\partial y_i}{\partial u_j}$. This choice of scale has, in fact, been widely used in linear least-squares problems as a device for improving the numerical aspects of computing procedures." [45]

where '$*$' denotes scaling and

$$A_{ij}^* = \frac{A_{ij}}{\sqrt{A_{ii}}\sqrt{A_{jj}}} \qquad (2.72)$$

$$g_i^* = \frac{g_i}{\sqrt{A_{ii}}} \qquad (2.73)$$

The operations above transfer the stiffness matrix $\mathbf{J}$ into a scaled symmetric square matrix $\mathbf{A}^*$.

Instead of solving the equation

$$\mathbf{A}^* \Delta \mathbf{u}_k^* = \mathbf{g}^*$$

the following equation is solved for $\Delta \mathbf{u}_k^*$:

$$(\mathbf{A}^* + \lambda_k \mathbf{I}) \Delta \mathbf{u}_k^* = \mathbf{g}^* \qquad (2.74)$$

$$\rightsquigarrow \quad \Delta \mathbf{u}_k^* = (\mathbf{A}^* + \lambda_k \mathbf{I})^{-1} \mathbf{g}^* \qquad (2.75)$$

where $\mathbf{I}$ is an identity matrix and λ_k is a scalar added to the diagonal of the matrix $\mathbf{A}^*$. Thereafter, the correction of the solution is obtained by scaling back $\Delta \mathbf{u}_k^*$:

$$\Delta \mathbf{u}_k^* = (\Delta u_i^*)_k \qquad (2.76)$$

$$\Delta u_i = \frac{\Delta u_i^*}{\sqrt{A_{ii}}} \qquad (2.77)$$

Finally, eq. (2.65) and $|\mathbf{y}_{k+1}|$ are computed.

It is noted that the operations of transforming $\mathbf{J}$ to $\mathbf{A}^*$ are found not to have noticeable effects to the solution. Thus, they, along with eq. (2.77), are not activated in SPARC.

According to [45], this algorithm performs like the gradient method[16] when a large value of λ is used. When a very small value of λ is used, this algorithm is the same as the Newton's method. Therefore, it performs an interpolation between the two methods by adjusting λ [45]. Another numerical benefit of this operation (eq. 2.74) is that $\mathbf{A}^*$ becomes better conditioned [45]. Before Marquardt (1963) proposed the above algorithm, Levenberg (1944, [35]) had recommended that adding some quantity to the diagonal to minimize $|\mathbf{y}|$ results in convergence similar to that in the gradient method. Any algorithms for optimization problems using the form of eq.

---

[16]Also known as the steepest-descent method. The gradient method solves the system of equation $y_k(\mathbf{u}) = 0$ with n unknowns by minimizing $\Phi = \sum_k^n y_k(\mathbf{u})^2$. The correction $\Delta u_j = -a\frac{\partial \Phi}{\partial u_j}$ is evaluated, where $\frac{\partial \Phi}{\partial u_j} = 2\sum_k^n y_k \frac{\partial y_k}{\partial u_j}$ and a is a properly defined constant.

(2.74) are therefore recognized as variations of the Levenberg-Marquardt algorithm in the literature.

In [45], a strategy for the determination of λ is proposed to accelerate slow convergence assuring

$$|\mathbf{y}_{k+1}| \leq |\mathbf{y}_k| \qquad (2.78)$$

According to this strategy, instead of using eq. (2.65), the iterates are updated by:

$$\mathbf{u}_{k+1} = \mathbf{u}_k + K\Delta\mathbf{u}_k \qquad (2.79)$$

K is determined by the angle between the correction $\Delta\mathbf{u}$ and the residual $\mathbf{g}$:[17]

$$\cos\vartheta = \left(\frac{(\Delta\mathbf{u})^\mathrm{T}\mathbf{g}}{|\Delta\mathbf{u}||\mathbf{g}|} \right) \qquad (2.80)$$

It is suggested that "any proper method must result in a correction vector whose direction is within 90° of the negative gradient of $|\mathbf{y}|$". Hence, a criterion angle ϑ_0 is selected satisfying $\vartheta_0 < \pi/2$. Here $\vartheta_0 = \pi/4$ is adopted as suggested. As a result, in case a large value of λ is required, increase λ and let $K = 1$ as long as:

$$\vartheta \geq \vartheta_0 \qquad (2.81)$$

Otherwise, stop increasing λ, but instead, decrease K until eq. (2.78) is fulfilled.

The strategy of choosing λ_k is summarized as follows [45] : $\mathbf{u}_k$, $|\mathbf{y}_k|$, and λ_{k-1} in a new iteration k are the end products of the previous iteration $k-1$. (In the initial step, assign a reasonable initial guess $\mathbf{u}_1$, and a small value for λ_0; then compute $\mathbf{y}(\mathbf{u}_1)$.)

1. Let $K = 1$ and choose a value for ν with $\nu > 1$. ν is a constant to scale λ.

2. Compute $\mathbf{J}_k$ using $\mathbf{u}_k$.

3. Take two trial runs with the two values (1) λ_{k-1}/ν and (2) λ_{k-1} and compute the corresponding errors $|\mathbf{y}(\lambda_{k-1})|$ and $|\mathbf{y}(\lambda_{k-1}/\nu)|$, respectively. If $\lambda_{k-1}/\nu \leq \eta$, stop the down-scaling of λ. η is a small value which leads to the convergence similar to the Newton's method. Note that, in a trial run, $\mathbf{y}(\lambda)$ is obtained by the following process:

   (a) λ is used to compute $\Delta\mathbf{u}$ using eq. (2.74) and eq. (2.77).
   (b) $\mathbf{u}_{k+1}$ are computed using eq. (2.79).
   (c) Let $\mathbf{y}(\lambda) = \mathbf{y}(\mathbf{u}_{k+1}) = \mathbf{y}_{k+1}$.

---

[17]In [45], the symbol γ is used for ϑ. However, ϑ is used in this work for the consistent of symbols.

Here, instead of writing $\mathbf{y}(\mathbf{u}_{k+1})$, $\mathbf{y}(\lambda)$ is used to emphasize that the improved solution $\mathbf{u}_{k+1}$ is computed in a trial run using λ.

4. The trial satisfying (eq. 2.78) is accepted:
   (a) If $|\mathbf{y}(\lambda_{k-1}/\nu)| \leq |\mathbf{y}_k|$, let $\lambda_k = \lambda_{k-1}/\nu$.
   (b) If $|\mathbf{y}(\lambda_{k-1}/\nu)| > |\mathbf{y}_k|$ and $|\mathbf{y}(\lambda_{k-1})| \leq |\mathbf{y}_k|$, let $\lambda_k = \lambda_{k-1}$.
   (c) If $|\mathbf{y}(\lambda_{k-1}/\nu)| > |\mathbf{y}_k|$ and $|\mathbf{y}(\lambda_{k-1})| > |\mathbf{y}_k|$, do the following:

       i. Compute $\Delta\mathbf{u}_k$ using the new trial $\lambda_{k-1}\nu^w$, and then compute the angle ϑ in eq. (2.80).
       ii. If $\vartheta \geq \vartheta_0$, then

           • Let $K_i = 1$ and compute $\mathbf{u}_{k+1}$ using eq. (2.79).
           • If $|\mathbf{y}(\lambda_{k-1}\nu^w)| \leq \mathbf{y}_k$, let $\lambda_k = \lambda_i\nu^w$;
             otherwise, increase w and repeat step 4.(c).

           Otherwise,
             As long as $|\mathbf{y}(\lambda_{k-1}\nu^w)| > \mathbf{y}_k$, decrease K and compute $\mathbf{u}_{k+1}$ using eq. (2.79). Whenever $|\mathbf{y}(\lambda_{k-1}\nu^w)| \leq \mathbf{y}_k$, let $\lambda_k = \lambda_i\nu^w$.

When the value of λ becomes too large, the corrections are nearly zero. In such situation, eq. (2.78) is fulfilled and convergence is not possible.

Discussion:
Based on the author's experience in using Marquardt's strategy for choosing λ, it has been found out:

• In the above strategy 4.(c), for $\vartheta \leq \vartheta_0$ convergence is not possible.
• When it fails to converge, re-starting the algorithm and choosing another μ might lead to convergence.
• The upper bound for λ is set as 10^{50} since convergence stops for such large value of λ; and the lower bound for λ is set to 10^{-14}.

2.7.4 Arc-length method

In this section, the arc-length method and pseudo-arc-length method are detailed. They are alternatives to Marquardt's method. In arc-length methods, an addition variable is added to a system. The arc-length method uses arc-length parameterization which provides an arc-length condition serving as an extra equation, whereas the pseudo-arc-length method uses an equation of a plane as the extra equation to the system of equations. The rationale and implementation of the arc-length methods are detailed herein. A calculation example of the arc-length methods is given in Appendix K.

2.7.4.1   Selection of an extra variable

The system of equations to be solved in SPARC is $\mathbf{y}(\mathbf{u}^t) = \mathbf{0}$ (eq. 2.51) with $\mathbf{y} \in \mathbb{R}^{d^{\text{of}}}$ consisting of equilibrium equations and $\mathbf{u}^t \in \mathbb{R}^{d^{\text{of}}}$ consisting of the unknown velocity components. However, the explanation of the arc-length methods is carried out using $\tilde{\mathbf{y}}(\mathbf{w}^{t+\Delta t}) = \mathbf{0}$ (eq. 2.53), instead of $\mathbf{y}(\mathbf{u}^t) = \mathbf{0}$ (eq. 2.51) (see **Remark 1** in Section 2.6.2 and **Remark 2** in Section 2.6.3), because the positions of the particles are the variables used in the arc-length parameterization (as will be shown later).

A parameter which has an influence on the system is chosen as an extra variable. For example, in the simulation of a triaxial test or an oedometer test, the system changes subject to the movement of top plate. In such cases, the position of the top plate at a later time $t + \Delta t$, $x_e^{t+\Delta t}$, can be selected as the extra variable in the arc-length method.

The system of equations adding an extra parameter as an unknown now reads:

$$\tilde{\mathbf{y}} = \tilde{\mathbf{y}}(\mathbf{w}^{t+\Delta t}, x_e^{t+\Delta t}) = \mathbf{0} \tag{2.82}$$

Let

$$\gamma^{t+\Delta t} = \begin{bmatrix} \mathbf{w}^{t+\Delta t} \\ x_e^{t+\Delta t} \end{bmatrix} \tag{2.83}$$

the system has $d^{\text{of}} + 1$ unknowns ($\gamma^{t+\Delta t} \in \mathbb{R}^{d^{\text{of}}+1}$). An additional equation will be obtained by applying arc-length parameterization.

2.7.4.2   Definition of *Parameterization according to arc length*

Let $[\alpha, \beta] \longrightarrow \mathbb{R}^N : L \longmapsto \gamma(L)$ be a parameterized curve. γ is said to be *parameterized according to arc length* L if the following two conditions are fulfilled [55]:

(A) $\forall L : |\dot{\gamma}| = 1$
(B) $L \in [\alpha, \beta]$

Condition (A) says that the tangent vector $\dot{\gamma}$ must have a length of 1 (Fig. 2.6), and condition (B) means the length of the curve is $L = \beta - \alpha$ when $|\dot{\gamma}| = 1$. Both conditions are valid simultaneously. A detailed explanation of the arc-length parameterization is provided in Appendix H.

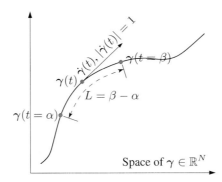

Figure 2.6: Taking $|\dot{\gamma}(t)| = 1$ during time $t \in [\alpha, \beta]$ results in $L = \beta - \alpha = \Delta t$ and vice versa.

2.7.4.3 Applying arc-length parameterization

Now, we apply the arc-length parameterization to provide an additional equation to the equations of system (eq. 2.82). When γ consists of the positions of particles, the tangent vector $\dot{\gamma}^t$ at time t in the arc-length parameterization consists of the velocities $\boldsymbol{\nu}^t$ of particles:

$$\dot{\gamma}^t = \begin{bmatrix} \frac{\mathbf{w}^{t+\Delta t} - \mathbf{w}^t}{\Delta t} \\ \frac{x_e^{t+\Delta t} - x_e^t}{\Delta t} \end{bmatrix} = \begin{bmatrix} \mathbf{u}^t \\ v_e^t \end{bmatrix} = \boldsymbol{\nu}^t \tag{2.84}$$

and leads to $\gamma^{t+\Delta t} = \gamma^t + \dot{\gamma}^t \Delta t$

The arc-length parameterization restricts the length of the tangent vector to 1:

$$|\dot{\gamma}^t| = |\boldsymbol{\nu}^t| = 1 \tag{2.85}$$

so that the distance L in the solution space (space of γ, see Fig. 2.6) between positions γ^t and $\gamma^{t+\Delta t}$ ($\gamma(t = \beta)$, and $\gamma(t = \alpha)$ in Fig. 2.6, respectively) equals Δt. This equation (eq. 2.85) serves as an additional equation to the system and has been termed normalization condition (Wubs 2009 [56]), arc-length condition (Higgins 2007 [24]) or scalar normalization (Keller 1986 [26]).

As a result, the new system of equations formulated using the unknown positions $\gamma^{t+\Delta t}$ reads:

$$\begin{cases} \tilde{\boldsymbol{y}} = \tilde{\boldsymbol{y}}(\gamma^{t+\Delta t}) = \mathbf{0} \\ |\dot{\gamma}^t| = 1 \end{cases} \tag{2.86}$$

And the system formulated using unknown velocities $\boldsymbol{\nu}^t$ (eq. 2.84) reads:

$$
\begin{cases}
\widetilde{\mathbf{y}} = \widetilde{\mathbf{y}}(\boldsymbol{\nu}^t) = \mathbf{0} & \text{(equilibrium equations)} \\
|\boldsymbol{\nu}^t| = 1 & \text{(arc-length condition)}
\end{cases}
\tag{2.87}
$$

Eq. (2.87) is the system of equations after applying the arc-length method. It is then solved using Newton's method.

Due to the massive usage of symbols, comparisons of symbols (before and after an extra parameter is added to the system of equations) are provided in Table 2.3.

Table 2.3: List of symbols used in arc-length methods.

Velocity		Position	
v	**velocity** vector of *one* particle	x	**position** vector of *one* particle
u	a vector consisting of merely unknown **velocity** components, eq. (2.50)	w	a vector consisting of merely unknown **position** components, eq. (2.52)
$\boldsymbol{\nu} = \begin{bmatrix} \mathbf{u} \\ v_e \end{bmatrix}$	a vector consisting of unknown **velocity** components plus an extra parameter v_e, eq. (2.84)	$\boldsymbol{\gamma} = \begin{bmatrix} \mathbf{w} \\ x_e \end{bmatrix}$	a vector consisting of merely unknown **position** components plus an extra parameter x_e, eq. (2.83)
$\mathbf{y}(\mathbf{u})$	System of equation using $\mathbf{u}$, eq. (2.51)	$\mathcal{Y}(\mathbf{w})$	System of equation using $\mathbf{w}$, eq. (2.53)
$\widetilde{\mathbf{y}}(\boldsymbol{\nu})$	System of equation using an extra parameter, eq. (2.87)	$\widetilde{\mathcal{Y}}(\boldsymbol{\gamma})$	System of equation using an extra parameter, eq. (2.86)

Since the unknowns are velocities $\boldsymbol{\nu}^t$, the explanation of the arc length must be carried out in the space of positions $\boldsymbol{\gamma}$, as illustrated in Fig. 2.7. It can be seen in Fig. 2.7 that the arc-length method is to find the solution $\boldsymbol{\gamma}^{t+\Delta t}$ which is an arc length $L\ (= \Delta t)$ away from the previous one ($\boldsymbol{\gamma}^t$) by applying the arc-length condition $|\dot{\boldsymbol{\gamma}}^t| = |\boldsymbol{\nu}^t| = 1$. Naturally, the arc length must be small enough compared to the radius of curvature of the solution path [26].

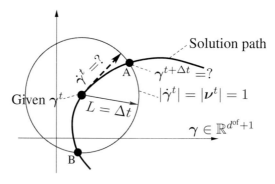

Figure 2.7: Illustration of the arc-length method.

Fig. 2.7 also shows that the arc-length method searches a solution on the circumference of the circle $|\boldsymbol{\nu}^t| = 1$. Accordingly, the solution can be at either point A or B (this can also be seen in the calculation example in Appendix K). As a result, the arc-length method does not provide a means for preserving searching directions of the solutions in Newton's iteration. The second power in $|\boldsymbol{\nu}^t| = 1$ permits the solutions to change their signs in the Newton's iteration. In a displacement-controlled simulation, for example, a change in the sign of v_e^t indicates that the loading direction has been changed. This also means the Jacobian matrix is leading the solution towards a searching direction opposite to the correct one. In this case, when the updated $v_e^{t,k+1}$ in Newton's iteration No. $k + 1$ changes its sign, it must be re-calculated using the negative values of Jacobian matrix (of the system in eq. 2.87). For complex boundary conditions, Keller [26] suggested the following relation to be fulfilled:

$$\left[\boldsymbol{\nu}^{t,\, k+1}\right]^{\mathrm{T}}\left[\boldsymbol{\nu}^{t-\Delta t,\, 0}\right] > 0 \tag{2.88}$$

As illustrated in Fig. 2.8, this inequality means that the angle between the updated solution vector $\boldsymbol{\nu}^{t,\, k+1}$ (in Newton's iteration No. $k + 1$, a column vector) and the solution vector $\boldsymbol{\nu}^{t-\Delta t,\, 0}$ (determined at the previous time step $t - \Delta t$) is less than $90°$. The searching direction is turned to an opposite direction if this inequality is violated.

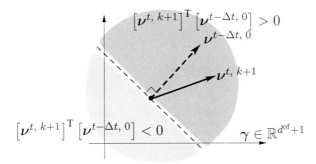

Figure 2.8: Illustration of the criterion (eq. 2.88) proposed by Keller [26] in Newton's itcration. k is the iteration counter.

2.7.4.4 Pseudo-arc-length method

The idea of the arc-length method is to search the solution on the circumference of the circle $|\dot{\boldsymbol{\gamma}}| = 1$. Another idea is to search the solution on a plane $\boldsymbol{\gamma} \cdot \mathbf{n} = \mathbf{0}$ in space $\boldsymbol{\gamma} \in \mathbb{R}^{d^{\mathrm{of}}+1}$, i.e. the pseudo-arc-length method [26]. It provides a means for preserving searching directions [26]. As illustrated in Fig. 2.9, the plane is constructed at $\boldsymbol{\gamma}_A$ with normal vector $\mathbf{n}$ on the plane.

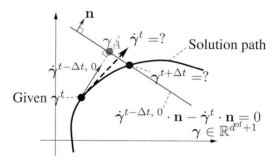

Figure 2.9: Illustration of the pseudo-arc-length method.

γ_A is away from $\gamma^{t+\Delta t}$ along the orientation of $\dot{\gamma}^{t-\Delta t,\,0}$, which has been determined at time step $t-\Delta t$. The unit normal vector $\mathbf{n}$ is selected parallel to $\dot{\gamma}^{t-\Delta t,\,0}$:

$$\mathbf{n} = \frac{\dot{\gamma}^{t-\Delta t,\,0}}{|\dot{\gamma}^{t-\Delta t,\,0}|} \tag{2.89}$$

The equation of the plane reads:

$$\dot{\gamma}^{t-\Delta t,\,0} \cdot \mathbf{n} - \dot{\gamma}^t \cdot \mathbf{n} = 0 \tag{2.90}$$

The derivation of eq. (2.90) is given in Appendix I.

With eq. (2.84), the extra equation provided by the plane, parameterized using velocity reads:

$$\boldsymbol{\nu}^{t-\Delta t,\,0} \cdot \mathbf{n} - \boldsymbol{\nu}^t \cdot \mathbf{n} = 0 \tag{2.91}$$

As a result, the system of equation in SPARC using pseudo-arc-length method is

$$\begin{cases} \widetilde{\mathbf{y}} = \widetilde{\mathbf{y}}(\boldsymbol{\nu}^t) = \mathbf{0} & \text{(equilibrium equations)} \\ \boldsymbol{\nu}^{t-\Delta t,\,0} \cdot \mathbf{n} - \boldsymbol{\nu}^t \cdot \mathbf{n} = 0 & \text{(pseudo-arc-length condition)} \end{cases} \tag{2.92}$$

A calculation example of the pseudo-arc-length method is given in Appendix K. The pseudo-arc-length method is used throughout this work.

2.7.4.5 Some remarks

The convergence is sometimes poor due to oscillations of the solutions in the Newton's iteration. In this case, it helps to add a reduction factor $f_r \in [0, 1]$ while updating the iterates in Newton's iteration:

$$\Delta \boldsymbol{\nu} = -\mathbf{J}^{-1} \begin{bmatrix} \widetilde{\mathbf{y}}\left(\boldsymbol{\nu}^{t,\,k}\right) \\ \boldsymbol{\nu}^{t-\Delta t,\,0} \cdot \mathbf{n} - \boldsymbol{\nu}^{t,\,k} \cdot \mathbf{n} \end{bmatrix} \tag{2.93}$$

$$\boldsymbol{\nu}^{t,\,k+1} = \boldsymbol{\nu}^{t,\,k} + f_r \Delta \boldsymbol{\nu} \tag{2.94}$$

where $\Delta \boldsymbol{\nu}$ is the correction applied to the variables (iterates) in Newton's iteration; $\mathbf{J}$ is the Jacobian matrix of the new system of equations (eq. 2.87). With the proper choice of f_r, we can achieve convergence, which however, can be very slow. A smaller value of f_r leads to a lower rate of convergence. Note that an unnecessary large number of iterations is required if f_r is too small.

When convergence can not be achieved by decreasing f_r, the author suggests decreasing the time increment Δt by a reduction factor f_t and restarting the same simulation step, e.g. $f_t = 0.5$.

2.7.5 Time integration scheme of stress

Inaccurate time integration of stress $\mathbf{T}^{t+\Delta t}$ often results in numerical instability in highly nonlinear problems. Since barodesy is a highly nonlinear constitutive model, an accurate time integration scheme shall be adopted. In SPARC, the substep method coupled with the fourth-order RUNGE-KUTTA method is employed to predict the stress $\mathbf{T}^{t+\Delta t}$ incrementally.

2.7.5.1 Substeps

Given a differential equation

$$f = \frac{\partial y}{\partial t} \tag{2.95}$$

and the value $y_0 = y(t = t_0)$, the forward EULER method (first-order approximation) approximates the solution as [33]

$$y_{i+1} = y_i + f_i \, \Delta t + \mathcal{O}(\Delta t^2) \tag{2.96}$$

$$f_i = f(t_i, \; y_i) \tag{2.97}$$

where Δt is the step size. The integration using forward EULER method is shown in Fig. 2.10 marked by a thin gray line. The idea of substepping is to divide the step size Δt into smaller increments $h = \Delta t / n$, where n is the number of substeps. The time integration of stress is proceeded incrementally using n substeps, as shown in

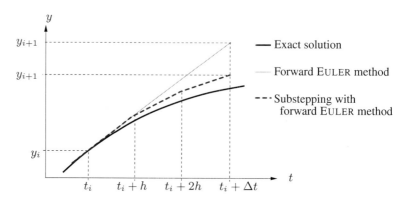

Figure 2.10: Example of using 3 substeps in the time integration from t_i to $t_i + \Delta t$, given y_i, t_i and the differential equation $f = \partial y / \partial t$. The forward EULER method is used as an example.

Fig. 2.10 marked as dashed line. Obviously, the result $y_{i+1} = y(t_i + \Delta t)$ determined by using 3 substeps is more accurate.

In SPARC, substeps are used in the time integration of the stress $\mathbf{T}^{t+\Delta t}$. The following code contains the implemented substep method in SPARC. For simplicity, the forward EULER method ($\mathbf{T}^{t+\Delta t} = \mathbf{T}^t + \dot{\mathbf{T}}^t \Delta t$) is used in the following code. Given $\mathbf{D}$, $\mathbf{W}$, $\mathbf{T}^t$, e^t, and the constitutive model barodesy $\mathcal{B}$ (eq. 2.36), the stress state $\mathbf{T}^{t+\Delta t}$ is evaluated incrementally in n substeps.

$$h = \Delta t / n \tag{2.98}$$

$$\mathbf{T}^{\mathsf{j}=1} = \mathbf{T}^t \tag{2.99}$$

$$e^{\mathsf{j}=1} = e^t \tag{2.100}$$

```
for j = 1:n
```

$$\mathring{\mathbf{T}}^{\mathsf{j}} = \mathcal{B}\left(\mathbf{T}^{\mathsf{j}}, e^{\mathsf{j}}, \mathbf{D}\right) \tag{2.101}$$

$$\dot{\mathbf{T}}^{\mathsf{j}} = \mathring{\mathbf{T}}^{\mathsf{j}} - \mathbf{T}^{\mathsf{j}} \mathbf{W} + \mathbf{W} \mathbf{T}^{\mathsf{j}} \tag{2.102}$$

$$\mathbf{T}^{\mathsf{j}} = \mathbf{T}^{\mathsf{j}} + \dot{\mathbf{T}}^{\mathsf{j}} h \tag{2.103}$$

$$e^{\mathsf{j}} = e^{\mathsf{j}} + \left(1 + e^{\mathsf{j}}\right) \operatorname{tr}\left(\mathbf{D}\right) h \tag{2.104}$$

```
end
```

$$\dot{e}^{t+\Delta t} = (e^{\mathsf{j}=\mathsf{n}} - e^t)/\Delta t \tag{2.105}$$

$$\dot{\mathbf{T}}^t = (\mathbf{T}^{\mathsf{j}=\mathsf{n}} - \mathbf{T}^t)/\Delta t \tag{2.106}$$

This method increases the accuracy of the time integration without increasing the total number of simulation steps.[18]

For $n = 1$, no substeps are used. The function $\mathcal{B}$ in eq. (2.101) denotes the constitutive model barodesy, and the Jaumann-Zaremba's objective stress rate is applied in eq. (2.102). The time integrations in eq. (2.103) and eq. (2.104) apply the forward EULER method. The void ratio e must be updated in the substeps as well (eq. 2.104), as it is taken into account in barodesy.

The results of the three-dimensional simulations of oedometer tests[19] show that the stress-strain curves obtained using substeps agree well with the simulation result of an element test[20], as shown in Fig 2.11. This proves that using substeps results in more accurate time integration of the stress.

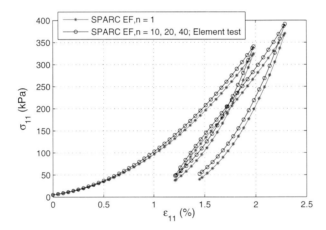

Figure 2.11: Stress-strain curves of the oedometer test using the forward EULER method (EF) and various numbers of substeps n. Note that the curves of the simulations adopting $n = 10, 20, 40$ substeps overlap with the element test curve (the discrepancies are not noticeable in the used scale). The discrepancy between the curve from the simulation without using substeps and the one from the element test is significant.

---

[18]Increasing the total number of simulation steps can significantly increase the computing time, as the construction of the Jacobian matrix in the Newton's iteration is computationally demanding.

[19]Degrees of freedom = 225; this corresponds to 125 particles. (see Section 4.1)

[20]Namely the stress-strain curve, given the stretching tensor $\mathbf{D} = \begin{bmatrix} \pm 1 & 0 & 0 \\ 0 & 0 & 0 \\ 0 & 0 & 0 \end{bmatrix}$, with $D_{11} = -1$

for oedometric compression and $D_{11} = +1$ for extension. Totally, 23358 steps are used to obtain this curve and the fourth-order RUNGE-KUTTA method is adopted for time integration.

2.7.5.2   Fourth-order RUNGE-KUTTA method

Higher-order methods such as HEUN's method (second order) [9, 33], RICHARD-SON's (half-step) extrapolation (second order)[21], and RUNGE–KUTTA methods (n-th order) [9, 33], to name a few, can significantly increase the accuracy of time integration in a time step. The fourth-order RUNGE-KUTTA method is implemented in SPARC.

The fourth-order RUNGE-KUTTA method approximates f similar to the forward EULER method:

$$y_{i+1} = y_i + K\,h + \mathcal{O}(h^5) \tag{2.107}$$

with K being evaluated as a weighted average slope:

$$K = \frac{k_1 + 2k_2 + 2k_3 + k_4}{6} \tag{2.108}$$

where

$$k_1 = f_i = f\left(t_i,\ y_i\right) \tag{2.109}$$

$$k_2 = f\left(t_i + \frac{h}{2},\ y_i + k_1\frac{h}{2}\right) \tag{2.110}$$

$$k_3 = f\left(t_i + \frac{h}{2},\ y_i + k_2\frac{h}{2}\right) \tag{2.111}$$

$$k_4 = f\left(t_i + h,\ y_i + k_3\,h\right) \tag{2.112}$$

k_1, computed at t_i, is identical to the slope f_i used in eq. (2.96). k_2, k_3 are evaluated at $t_i + h/2$. k_4 is evaluated at $t_i + h$.

The fourth-order RUNGE-KUTTA method is applied to compute a well-estimated rate $\overset{\circ}{\mathbf{T}}$ in eq. (2.101). Thus, in the substeps, we may simply replace eq. (2.101) by the following code. The computation procedure follows eq. (2.108) through eq. (2.112):

$$\dot{e} = (1 + e^{i})\,\mathrm{tr}\,\mathbf{D} \tag{2.113}$$
$$\%\ \text{Evaluate } \overset{\circ}{\mathbf{T}}_1$$

---

[21]It coincides with the second order Runge–Kutta method and has been used in [17] with adaptive time integration scheme for problems modeled with hypoplastic constitutive equations.

$$\overset{\circ}{\mathbf{T}}_1 = \mathcal{B}\left(\mathbf{T}^{\mathrm{i}}, e^{\mathrm{i}}, \mathbf{D}\right) \tag{2.114}$$

% Evaluate $\overset{\circ}{\mathbf{T}}_2$

$$\mathbf{T}_1 = \mathbf{T}^{\mathrm{i}} + \overset{\circ}{\mathbf{T}}_1\,\frac{h}{2} \quad ; \quad e_1 = e^{\mathrm{i}} + \dot{e}\,\frac{h}{2} \tag{2.115}$$

$$\overset{\circ}{\mathbf{T}}_2 = \mathcal{B}\left(\mathbf{T}_1, e_1, \mathbf{D}\right) \tag{2.116}$$

% Evaluate $\overset{\circ}{\mathbf{T}}_3$

$$\mathbf{T}_2 = \mathbf{T}^{\mathrm{i}} + \overset{\circ}{\mathbf{T}}_2\,\frac{h}{2} \quad ; \quad e_2 = e_1 \tag{2.117}$$

$$\overset{\circ}{\mathbf{T}}_3 = \mathcal{B}\left(\mathbf{T}_2, e_2, \mathbf{D}\right) \tag{2.118}$$

% Evaluate $\overset{\circ}{\mathbf{T}}_4$

$$\mathbf{T}_3 = \mathbf{T}^{\mathrm{i}} + \overset{\circ}{\mathbf{T}}_3\,h \quad ; \quad e_3 = e^{\mathrm{i}} + \dot{e}\,h \tag{2.119}$$

$$\overset{\circ}{\mathbf{T}}_4 = \mathcal{B}\left(\mathbf{T}_3, e_3, \mathbf{D}\right) \tag{2.120}$$

% Evaluate $\overset{\circ}{\mathbf{T}}^{\mathrm{i}}$

$$\overset{\circ}{\mathbf{T}}^{\mathrm{i}} = \frac{1}{6}\left(\overset{\circ}{\mathbf{T}}_1 + 2\overset{\circ}{\mathbf{T}}_2 + 2\overset{\circ}{\mathbf{T}}_3 + \overset{\circ}{\mathbf{T}}_4\right) \tag{2.121}$$

This code calculates $\overset{\circ}{\mathbf{T}}^{\mathrm{i}}$ as the weighted average rate of $\overset{\circ}{\mathbf{T}}_1$, $\overset{\circ}{\mathbf{T}}_2$, $\overset{\circ}{\mathbf{T}}_3$, and $\overset{\circ}{\mathbf{T}}_4$.

The results with various substep numbers are shown in Fig. 2.12. All curves are in good agreement with the element test curve.

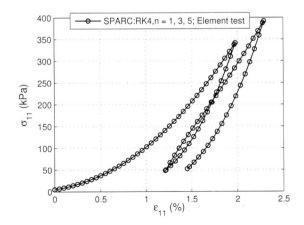

Figure 2.12: Stress-strain curves of the oedometer test using the fourth-order RUNGE-KUTTA method (RK4) and various numbers of substeps n. Note that all curves overlap with the element test curve (the discrepancies are not noticeable in the used scale).

In SPARC, the default number of substeps is 2 and the fourth-order RUNGE-KUTTA method is adopted for the time integration of stress. Compared with $n = 1$, using $n = 2$ increases the accuracy of the time integration by 2^4 times.[22]

Comparison: EULER method and RUNGE-KUTTA method with substeps

The close-up views of the two corners of the stress-strain curves (as show in Figs. 2.13a and 2.13b) reveal that the accuracy of the time integration using one substep ($n = 1$) coupled with the RUNGE-KUTTA method is much higher than that using 40 substeps ($n = 40$) with the EULER method. Note that the amount of computation for one substep using the RUNGE-KUTTA method is approximately comparable with four substeps using the EULER method.

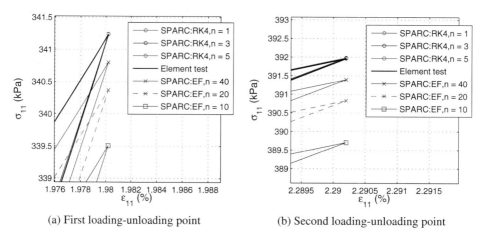

(a) First loading-unloading point   (b) Second loading-unloading point

Figure 2.13: Comparisons of results using the fourth-order RUNGE-KUTTA method (RK4) and the forward EULER method (EF) in various numbers of substeps. The close-up views of of the stress-strain curves at the two loading-unloading points in Fig. 2.11 are shown in (a) and (b). Note that discrepancies between the EF curves and the element test curve can be observed in these close-up views, whereas all RK4 curves overlap with the element test curve in the used scale.

---

[22]The RUNGE-KUTTA method is accurate of the order h^4, $n = 2$ leads to an improved accuracy of $\left(\frac{h}{2}\right)^4$.

2.8   Pressure boundary condition

Constant cell pressure is applied at the lateral boundary of a the triaxial test sample shown in Fig. 2.14. As the hydrostatic cell pressure p is perpendicular to the

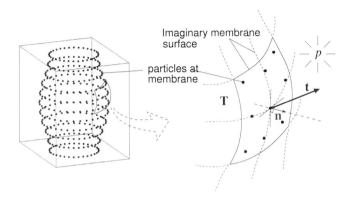

Figure 2.14: Equilibrium on boundaries subject to hydrostatic pressure p. $\mathbf{T}$ is the CAUCHY's stress tensor in a deformed specimen. $\mathbf{n}$ is the unit normal vector of the surface at the particle under evaluation and $\mathbf{t} = \mathbf{T}^{t+\Delta t}\mathbf{n}^{t+\Delta t}$ is the stress vector at that particle on the surface. The dots denote the surface particles. All parameters are referred to time $t + \Delta t$.

specimen surface, the stress vector $\mathbf{t}$ $(= \mathbf{T}^{t+\Delta t}\mathbf{n}^{t+\Delta t})$ on the surface with unit normal vector $\mathbf{n}^{t+\Delta t}$ must equal the stress $(-p)\mathbf{n}^{t+\Delta t}$ resulting from p. Therefore, the equilibrium condition for hydrostatic pressure boundaries reads

$$\mathbf{T}^{t+\Delta t}\mathbf{n}^{t+\Delta t} - (-p)\mathbf{n}^{t+\Delta t} = \mathbf{0} \qquad (2.122)$$

The evaluation of the unit normal vectors at surface particles is documented in Appendix D.

2.9   Parallelization

The parallelization of computations is conducted in SPARC for the following two computations.

1. Computation of Jacobian matrix:
   Each column in the Jacobian matrix is computed in one individual central processing unit (CPU).

2. Inversion of Jacobian matrix: When multi-cores are available, Matlab automatically makes use of the multi-cores while evaluating the correction of the iterates $-\left[y_{i_e,i_u}\right]_k^{-1} y_{i_e}(\mathbf{x}_k)$ in eq. (2.61). The stiffness matrix is stored as a sparse matrix to accelerate the computation. The sparsity of the Jacobian matrix and the computing efficiency using sparse matrix are documented in Appendix E.

Chapter 3

Parameter Study

In this chapter, the following issues are investigated and the findings are detailed:

- The configuration of particles.
- The neighbor search method.
- The support size (i.e. number of neighbors).
- The order of polynomials.
- The numerical solver.

Although the results may be case dependent, they provide information of the performance and limits of the current version of SPARC.

3.1   Benchmark

The simulation of a biaxial test under plane-strain condition is used as the benchmark in this chapter to investigate the aforementioned issues. The biaxial test is used for the following reasons:

- When frictionless boundaries (between the loading plates and the sample) and initial homogeneity in stress and void ratio are assumed, the deformation field (and hence the stress field) in the sample shall be homogeneous, i.e. an element test.
- Due to the fact that the material exhibits strain-softening, minor errors in the solution are accumulated with time, significant strain localization occurs when the specimen is axially compressed to some extent, resulting in the appearance of one or more shear bands.

The benchmark is detailed in the following sub-sections.

3.1.1   Particle configuration and boundary conditions

The biaxial specimen is discretized into 231 regularly arrayed particles. The configuration of particles is shown in Fig. 3.1a. The 2D space is demonstrated on the

y-z plane, with z being the axial direction. The line segments denote the kinematic degrees of freedom resulting from the prescribed boundary conditions illustrated in Fig. 3.1b.

Loading is displacement-controlled. The z-components in the velocity vectors of the particles on the loading plates are prescribed, with $v_z \neq 0$ and $v_z = 0$ for particles located on the top and bottom plates, respectively. As the interfaces are frictionless, v_y at these particles are unknowns, except for the one located in the middle of the top plate being fixed (i.e. $v_y = 0$).

To ensure that the material exhibits strain-softening, a dense sample with initial void ratio $e = 0.6324$ and cell pressure 100 kPa is prescribed. For surface particles, eq. (2.122) expresses equilibrium. In eq. (2.122), the normal vectors ($\mathbf{n}$) are evaluated on surface particles. The equilibrium for the remaining particles is described by eq. (2.2) with $\mathbf{g} = \mathbf{0}$[1].

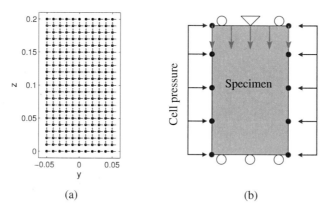

$$\text{(a)} \qquad\qquad\qquad\qquad \text{(b)}$$

Figure 3.1: (a) Configuration of particles. (b) Boundary conditions. In (a), the dots denote soft particles. The line segments at the particles denote the degrees of freedom in z- and y-direction. In (b), the blue solid dots denote schematically the particles to which the pressure boundary condition is applied.

3.1.2   Neighbor search and support size

The fixed-radius search method is used in the benchmark. The support size is determined by using $r = 1.7h$ with h being the distance between two adjacent particles, as illustrated in Fig. 3.2.

---

[1]The influence of gravity is negligible compared to the stress level in the simulation.

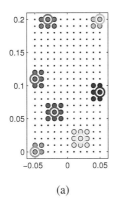 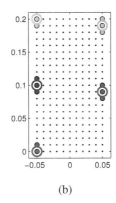

(a)   (b)

Figure 3.2: Examples of supports in the benchmark. The fixed-radius search method with $r = 1.7h$ is used, where h is the distance of two adjacent particles. The colored solid dots are neighbors of the circled particles. (a) Examples of supports used for the evaluation of spatial derivatives. (b) Examples of supports used for the evaluation of the normal vectors on the pressure boundaries. Note that the particles on the boundaries have fewer neighbors.

3.1.3 Polynomial order

First-order polynomials are used in the benchmark:

$$\hat{f} = a_1 x_1 + a_2 x_2 \tag{3.1}$$

3.2 Numerical solvers

The performance of the three numerical solvers detailed in section 2.7 (i.e. the Newton's method (NM), Levenberg-Marquardt method (L-M) with Marquardt's strategy for choosing λ and the pseudo-arc-length method (ARC)) is investigated.

3.2.1 Results from the benchmark

3.2.1.1 Convergence

The tolerance $\epsilon = 10^{-4}$ and the objective function value defined in eq. (2.63) are adopted in the solvers. The step size in NM and in ARC is reduced by a factor of 0.5 when the convergence is not possible (e.g. slow to no convergence, oscillation, or divergence)[2]. With L-M, reducing Δt usually does not improve convergence. It was found out that, in L-M, accepting the solution when the convergence stops is often followed by better convergence in later steps. However, the solution can be wrong in this case.

The stress-strain curves obtained using the three solvers are shown in Fig. 3.3. The curve obtained from element test simulation[3] is plotted for comparison. The solutions found by the three solvers are in good agreement with the element test case.

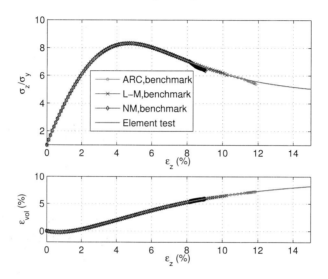

Figure 3.3: Simulation results in terms of stress-strain and $\varepsilon_{vol} - \varepsilon_z$ relations.

The convergence of the three solvers in terms of the numbers of iterations required to solve the system of equations is illustrated in Fig. 3.4. The stress-strain curves are plotted for comparison. L-M encounters convergence problems when the stress-strain curve comes close to the peak ($\varepsilon_a = 3.8\%$ and 4.0% with objective function values $\approx 10^{-2}$) and at the peak (at $\varepsilon_a = 4.7\% \ldots 5.1\%$ with objective function values $\approx 10^{-2}$). The convergence with objective function values $< 10^{-2}$ is not possible at these strains. After $\varepsilon_a = 8.7\%$, L-M can not find solutions with an

---

[2]A maximum number of iterations of 200 for ARC and NM are used to restart the solver with a reduced step size when slow to no convergence or oscillation occurs. For L-M, the solution is accepted without reducing step size. A maximum objective function value of 10^{-8} is used to restart the solver when the solution process diverges. Note that these values shall be adjusted based on the problem size.

[3]Namely, the stress-strain curve directly obtained from the constitutive model. It contains 2002 time steps and is obtained using an element-test simulation with initial void ratio $e = 0.6324$ under plane-strain condition, where $\mathbf{D} = \begin{bmatrix} \pm 1 & 0 & 0 \\ 0 & D_{22} & 0 \\ 0 & 0 & 0 \end{bmatrix}$ and initial $\mathbf{T} = \begin{bmatrix} -100 & 0 & 0 \\ 0 & -100 & 0 \\ 0 & 0 & -100 \end{bmatrix}$ kPa are prescribed and D_{22} is determined in each step by satisfying the condition $\dot{T}_{22} = 0$. The time integration scheme adopts three substeps coupled with the fourth-order Runge-Kutta method.

objective function value $< 10^2$ and the objective function value has been increased to 10^5 at $\varepsilon_a = 9.7\%$.

NM encounters convergence difficulties after the peak. A large number of iterations is often required to find the solution. With convergence difficulties, it reaches the axial strain around $\varepsilon_a = 7.7\%$, at which a shear band starts to form. Thereafter, the time step is decreased from $\Delta t = 10^{-3}$s down to $\Delta t = 10^{-7}$s in the next three steps. In the end, $\Delta t = 10^{-18}$s ($\varepsilon_a = 8.6\%$) is required for convergence.

ARC has no convergence problem at peak. In the softening region, it solves the system of equations with significantly fewer iterations than L-M and NM. The formation of shear bands occurs at $\varepsilon_a = 10.4\%$. Thereafter the solver still can solve the system without significantly reducing Δt until $\varepsilon_a = 11.2\%$

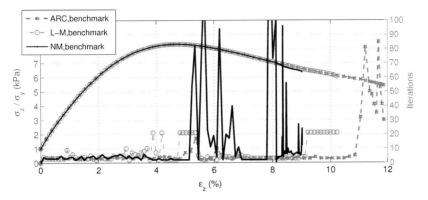

Figure 3.4: Comparison of the numbers of iterations required in the pseudo-arc-length method (ARC), Levenberg-Marquardt method (L-M), and Newton's method (NM). The corresponding stress-strain curves are plotted for comparison. The element test curve is the stress strain curve of the constitutive model barodesy.

3.2.1.2 Homogeneous fields and shear band formation

The boundary conditions applied in the benchmark result in homogeneous fields that correspond to element tests. The results obtained using ARC are taken to check the homogeneity of the results. In Fig. 3.5, the stress components σ_{zz} and σ_{yz} of $\mathbf{T}$ for all particles are plotted against ε_{zz}. The curves overlap with one another for $\varepsilon_a < 10.4\%$ indicating the homogeneity of the stress field. In addition, the homogeneity of deformation can be observed in the plots of the principal directions of incremental strain rates and the contours of the $|\mathbf{D}|$ field ($|\mathbf{D}| = \sqrt{\text{tr} (\mathbf{D}^{\mathsf{T}}\mathbf{D})}$), as shown in Fig. 3.6. These results evident that the elements used in the benchmark can model the homogeneous stress and deformation fields.

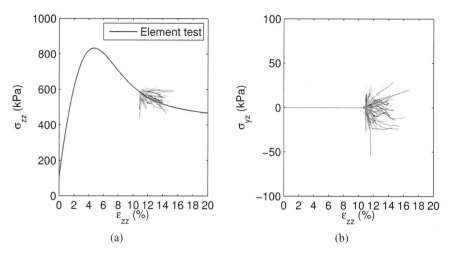

Figure 3.5: Stress-strain curves of all particles using solutions determined with the pseudo-arc-length method. (a) σ_{zz} vs. ε_{zz} curves. (b) σ_{yz} vs. ε_{zz} curves.

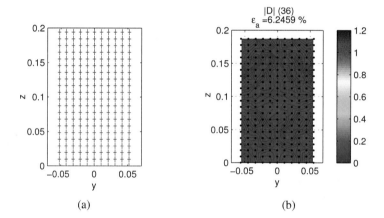

Figure 3.6: (a) Incremental principal strain directions and (b) Contours of the deformation field in terms of $|\mathbf{D}|$ at axial strain $\varepsilon_a = 2.92\%$ (as an example), using the pseudo-arc-length method. The length of the crosses denotes the magnitude of the principal strains.

The deformation field in terms of $|\mathbf{D}|$ shown in Fig. 3.7 indicates that the deformations significantly localize in one step from $\varepsilon_a = 10.26\%$ to 10.44% and a shear band forms. The velocity field in this step undergoes significant changes, as shown in Fig. 3.8, and the sample is separated into two rigid bodies. Some steps later, the deformation in the shear band increases, as shown in Fig. 3.9. It is noted that insignificant localization of the deformation occurs before the formation of the shear band (Fig. 3.7a). It results from the errors accumulated in the solutions. The shear band appears when the inhomogeneity is significant.

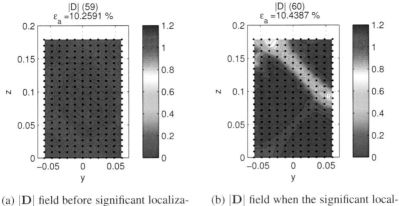

(a) $|\mathbf{D}|$ field before significant localization of deformations.

(b) $|\mathbf{D}|$ field when the significant localization of deformations begins.

Figure 3.7: Contours of the deformation fields using ARC, before and after the deformations significantly localize.

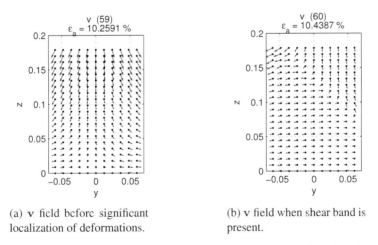

(a) $\mathbf{v}$ field before significant localization of deformations.

(b) $\mathbf{v}$ field when shear band is present.

Figure 3.8: Velocity field obtained using ARC, before and after the deformations significantly localize. Note that the solution fields change significantly in one step.

L-M and NM also yield homogeneous stress and incremental strain fields before strains significantly localize. NM can solve the system of equations when the strains localize at $\varepsilon_a = 7.7\%$ (requiring very small Δt to proceed after the formation of shear bands). However, the shear band pattern, as shown in Fig. 3.10, differs from the one determined by ARC. The shear band also forms in one simulation step. The solutions in this step also change significantly, as shown in Fig. 3.11.

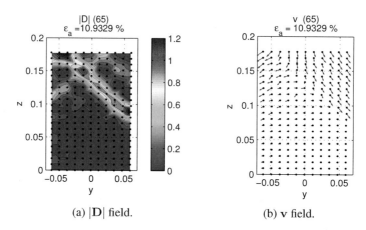

(a) |**D**| field.   (b) **v** field.

Figure 3.9: Illustration of (a) the shear band and (b) the corresponding velocity field.

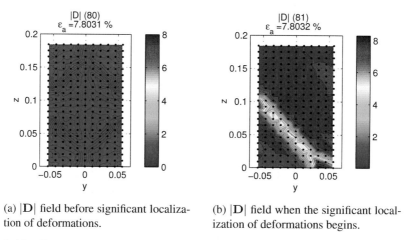

(a) |**D**| field before significant localiza-   (b) |**D**| field when the significant local-
tion of deformations.   ization of deformations begins.

Figure 3.10: Contours of the deformation fields using NM, before and after the deformations significantly localize. Note that this pattern is different from the one using ARC.

3.2.2   Role of a weak zone

A weak zone is implemented to enforce shearband. It is implemented by increasing the void ratio by 0.01 at two particles, as shown in Fig. 3.12. In Fig. 3.13, the numbers of iterations used to solve the system of equations are plotted along with the stress-strain curves.

The Levenberg-Marquardt method and the Newton's method do not converge after $\varepsilon_a = 4.2\%$ when the strains significantly localize. The Newton's method requires very small Δt to converge for some more steps and the Levenberg-Marquardt

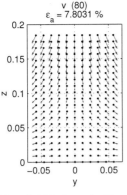
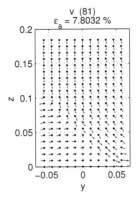

(a) **v** field before significant
localization of deformations.

(b) **v** field when shear band is
present.

Figure 3.11: Velocity field obtained using NM, before and after the deformations
significantly localize.

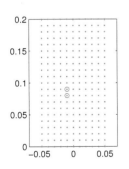
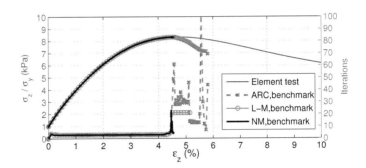

Figure 3.12:
Illustration of
the weak zone.
Void ratio of the
circled particles are
increased by 0.01.

Figure 3.13: Iterations required for solving the system of
equations. The stress-strain determined by the solvers are
plotted for comparison.

method could not find solution with sufficient small errors lower than the tolerance.
Although the arc-length method requires more iterations to find the solution, it can
solve the equation of systems up to $\varepsilon_a = 5.7\%$. The contours of deformations
around $\varepsilon_a = 4.2\%$ using solutions determined by the three solvers are shown in Fig.
3.14. The solutions (velocity fields) and the resulting shear band patterns determined
by the three solvers are very similar. The evolution of the shear bands is shown in
Appendix F. Due to the presence of a weak zone, the peak occurs earlier than the

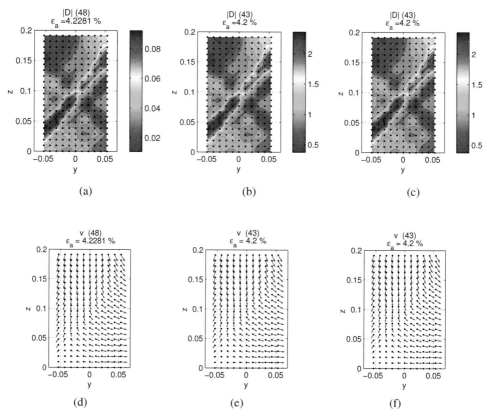

Figure 3.14: Contours of deformation fields around $\varepsilon_{zz} = 4.2\%$ using solutions (velocity field) found by the arc-length method (a and d), the Newton's method (b and e) and the Levenberg-Marquardt method (c and f).

one from element test. Besides, the shear band forms progressively, contrary to the formation of shear band that occurs in one step when no weak zone is present.

The stress-strain curves for all particles are plotted in Fig. 3.15, using solutions found by the arc-length method as an example. The stress components σ_{zz} and σ_{yz} of the Cauchy's stress tensor and the ε_{zz} component of the strain tensor are used for the plot. All curves, except for the two with higher void ratios, are overlapping until $\varepsilon_a = 4.2\%$ where the strains significantly localize. Fig. 3.15a indicates that, when the strains significantly localize, two tendencies of the curves' evolution are present. One tendency is that the stress-strain curve more or less follows the element test curve, with monotonic increases in ε_{zz}. Particles in the strain-localized band show this tendency, as shown in Fig. 3.16. The other tendency shows decrease in both ε_{zz} and σ_{zz} when significant strain localization is taking place. Particles in the rigid body areas show this tendency, as shown in Fig. 3.17.

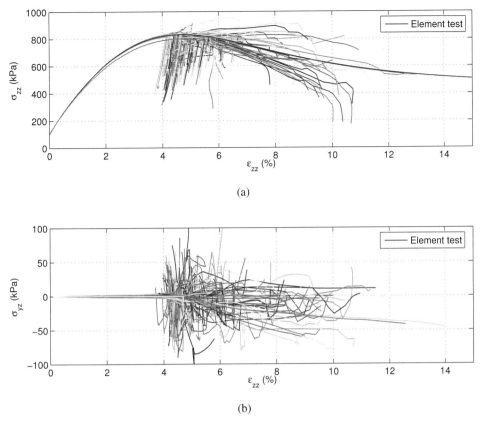

Figure 3.15: (a) Stress-strain curves using σ_{zz} and σ_{yz} components of stress tensors and ε_{zz} components of strain tensors.

3.2.3 Brief summary

The findings based on the benchmark are as follows:

- The arc-length method performs the best among the tested solvers. Levenberg-Marquardt method and the Newton method encounter convergence difficulties after the peak is reached, whereas the arc-length method finds solutions with significantly fewer steps.

- The benchmark can present the homogeneous deformation field and stress field before the appearance of the shear band. It can also model strain localization and the rigid-body movements separated by the shear band.

- Before the occurrence of the shear band, the solutions determined by all solvers are the same. Thereafter, the Newton's method and the arc-length method find different solutions (thus, different shear band patterns).

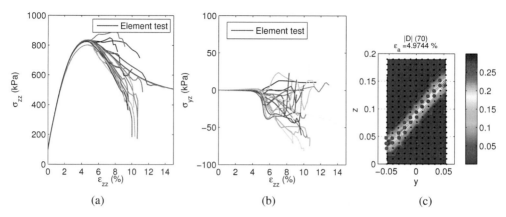

(a)   (b)   (c)

Figure 3.16: Stress-strain curves (a and b) of particles in the strain-localized region, marked as red in (c).

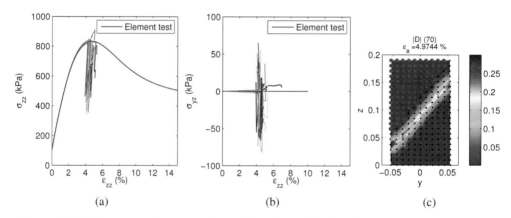

(a)   (b)   (c)

Figure 3.17: Stress-strain curves (a and b) of particles in the rigid blocks above the shear band, marked as red in (c).

3.3   Support size

The number of neighbors determined by the fixed-radius search method may vary from particle to particle. The particles on the boundaries have fewer neighbors than those in the interior of the body. The k-nn search method, on the contrary, assigns a fixed number of neighbors to every particle. In this section, the results using various support sizes and first- and second-order polynomials are presented.

3.3.1 First-order polynomials with fixed-radius search method

A simulation with larger support sizes than benchmark is carried out. The examples of supports are illustrated in Fig. 3.18a and Fig. 3.18b. The velocity field is smooth before the formation of shear band. It becomes less smooth as the strains localize in the shear bands, as shown in Fig. 3.18c and Fig. 3.18d.

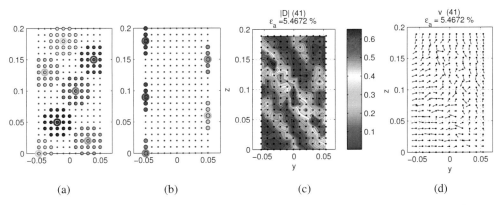

Figure 3.18: Examples of supports used for the evaluation of (a) spatial derivatives and (b) normal vectors on the pressure boundary. (c) Contours of $|\mathbf{D}|$ in step 41 (axial strain $\varepsilon_a = 5.4\%$) and (c) the corresponding velocity field determined using the arc-length method.

3.3.2 Second-order polynomials with fixed-radius search method

Second-order polynomials are used

$$\hat{f} = a_1 x_1 + a_2 x_2 + a_3 x_1 x_2 + a_4 x_1^2 + a_5 x_2^2 \tag{3.2}$$

It requires at least five particles in the support to determine the coefficients. The supports illustrated in Fig. 3.18a and Fig. 3.18b are used. It is reasonable to use one-dimensional second-order polynomials

$$\hat{f} = a_1 x_1 + a_1 x_1^2 \tag{3.3}$$

for the shape of the meridian of the sample and the evaluation of the normal vectors for the surface particles. The supports are illustrated in Fig. 3.18a and Fig. 3.18b.

The velocity field is homogeneous in the early stage of the simulation, e.g. at axial strain $\varepsilon_a = 2.6\%$ shown in Fig. 3.19a. The system can not be solved after $\varepsilon_a =$

3.6%, as convergence is not possible even with an extremely small step size $\Delta t = 10^{-10}$s. The velocity fields become less smooth with time, as shown in Fig. 3.19b.

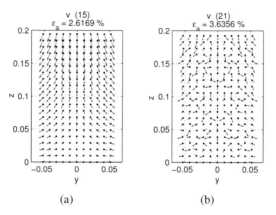

(a)   (b)

Figure 3.19: (a) The velocity field in step 15 (axial strain $\varepsilon_a = 2.6\%$) and (c) the velocity field in step 21 ($\varepsilon_a = 3.6\%$).

3.3.3   Second-order polynomials with k-nn search

Simulations with second-order polynomials using neighbor numbers $k = 9$ and $k = 15$ are tested and the results are presented in the following two subsections.

3.3.3.1   k = 9

The examples with $k = 9$ neighbors are illustrated in Fig. 3.20a. The support used for the determination of the normal vector (with one-dimensional second-order polynomials) on the pressure boundary is shown in Fig. 3.20b. The deformation fields using the determined solution are not homogeneous at all times. Fig. 3.21a, for example, illustrates the solution field at axial strain $\varepsilon_a = 0.63\%$. The resulting deformation field, as shown in Fig. 3.21b, reveals that deformations localize on the bottom of the specimen. The system can not be solved at axial strain $\varepsilon_a = 1.2\%$ and no convergence is possible even for a very small step size.

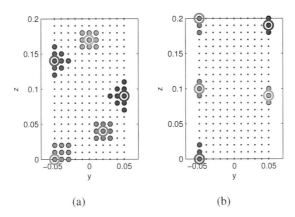

(a)   (b)

Figure 3.20: (a) Support with $k = 9$ neighbors in the k-nn search method. (b) Support with $k = 3$ neighbors in the k-nn search method.

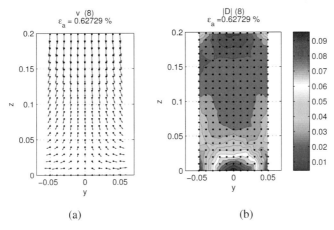

(a)   (b)

Figure 3.21: (a) Velocity field determined at axial strain $\varepsilon_a = 0.63\%$. (b) Contours of the corresponding deformation field.

3.3.3.2 k = 15

The examples with $k = 15$ neighbors used for the determination of the spatial derivatives are illustrated in Fig. 3.22a. The support used for the determination of the normal vector (with one-dimensional second-order polynomials) on the pressure boundary is shown in Fig. 3.22b. In the first 20 steps, the deformation fields are homogeneous, as shown in Fig. 3.23. The deformations begin to concentrate around the upper-right and lower-left corners of the sample with time, as shown in Fig. 3.24. Thereafter, significant strain localization takes place at the upper-right and lower-left corners (Fig. 3.25). The corresponding velocity field (Fig. 3.26) shows that a shear band, oriented from upper-right to lower-left, appears. The shear band is thick and separates two rigid bodies on the upper-left and lower-right corner. Convergence after $\varepsilon_a = 2.62\%$ is not possible. The width of shear band results from the larger number of neighbors and the second order polynomials being used.

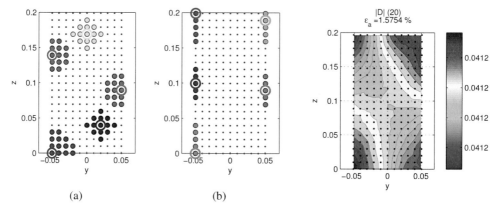

(a)   (b)

Figure 3.22: Examples of supports. (a) 15 neighbors for inner particles and (a) 5 neighbors for surface particles.

Figure 3.23: Contours of $|\mathbf{D}|$ at axial strain $\varepsilon_a = 1.58\%$

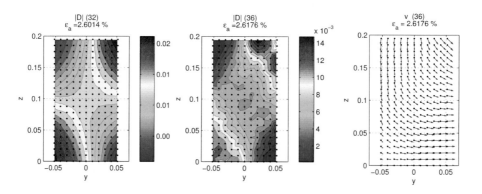

Figure 3.24: Contours of $|\mathbf{D}|$ at axial strain $\varepsilon_a = 2.60\%$

Figure 3.25: Contours of $|\mathbf{D}|$ at $\varepsilon_a = 2.62\%$

Figure 3.26: Velocity field determined at $\varepsilon_a = 2.62\%$

3.3.4 Brief summary

k-nn search assigns the same number of neighbors to each particle. Consequently, compared with particles not on the boundaries, the particles on the boundaries have neighbors which are in a further distance. In this section, it has been shown that:

1. In the simulations with first-order polynomials, using k-nn search leads to inhomogeneous deformations at axial strain $\varepsilon_a = 0.63\%$. However, the benchmark case can simulate both homogeneous deformation and the shear band formation. Thus, k-nn search is not superior to the fixed-radius search method.
2. With second-order polynomials, in the simulation using the k-nn search, the homogeneous deformation and thereafter shear band formation can be modeled. However, using the fixed-radius search results in inhomogeneous velocity fields.
3. The shear band appears at smaller axial strain in the simulation with second-order polynomials.

3.4 Irregular array of particles

Four simulations with the following elements are carried out and the results are presented in this section:

- First-order polynomials with fixed-radius search method
- Second-order polynomials with fixed-radius search method
- First-order polynomials with k-nn search
- Second-order polynomials with k-nn search

Note that the particle configurations in the above simulations are the same.

3.4.1 First-order polynomials with fixed-radius search method

The fixed-radius search method are tested herein. The results using radius $r = 0.012$ and $r = 0.0105$ are presented herein.

3.4.1.1 Radius r = 0.012 m

The supports determined by a radius $r = 0.012$ m are illustrated in Fig. 3.27. The deformation field is homogeneous with insignificant discrepancies present as indicated by the contours shown in Fig. 3.28. However, after $\varepsilon_a = 1.9\%$, no convergence is possible. Note that the pattern of the insignificant discrepancies is much less uniform compared to the one in Fig. 3.23, in which a regularly allocated particle configuration is used.

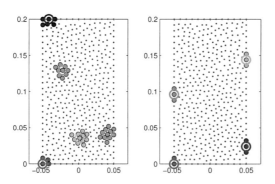

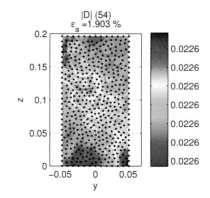

Figure 3.27: (a) Configuration of the particles and examples of supports determined using the fixed-radius search method. (b) Examples of supports used for the determination of the normal vectors on the pressure boundary.

Figure 3.28: Contours of the $|\mathbf{D}|$ field at axial strain $\varepsilon_a = 1.90\%$.

3.4.1.2   Radius $r = 0.0105$ m

Examples of supports determined by a smaller radius $r = 0.0105$ m are illustrated in Fig. 3.29.

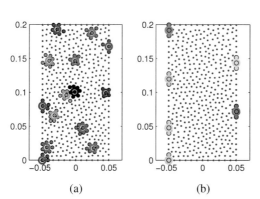

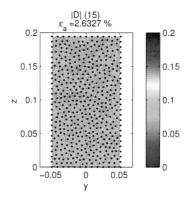

(a)   (b)

Figure 3.29: Examples of supports used for the evaluation of (a) spatial derivatives and (b) normal vectors on the pressure boundary.

Figure 3.30: Contours of the $|\mathbf{D}|$ field at axial strain $\varepsilon_a = 2.63\%$.

The deformation field is homogeneous until $\varepsilon_a = 2.63\%$, as shown in Fig. 3.30. Thereafter, the strains localize in some areas, as shown in Fig. 3.31 and Fig. 3.32, and the solver (ARC) can not find solutions after $\varepsilon_a = 3.25\%$ even with a very small Δt.

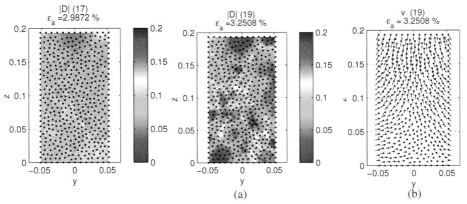

Figure 3.31: Contours of the |D| field at axial strain $\varepsilon_a = 2.99\%$.

Figure 3.32: (a) Contours of the deformation field at axial strain $\varepsilon_a = 3.25\%$ and (b) the corresponding velocity field.

The stress-strain curves (in terms of σ_{zz} and σ_{yz}) of all particles are shown in Fig. 3.33. All curves overlap with the element test curve, indicating that the stress field is homogeneous during the simulation. No solutions can be found by the arc-length method after $\varepsilon_a = 3.25\%$.

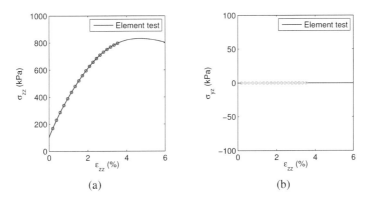

Figure 3.33: Stress-strain curves of all particles using solutions determined with the arc-length method. (a) σ_{zz} vs. ε_{zz} curves. (b) σ_{yz} vs. ε_{zz} curves.

3.4.2 Second order polynomials with fixed-radius search method

The supports used for the evaluation of the spatial derivatives and the normal vectors on the pressure boundary are shown in Fig. 3.34. The contours of the deformation field in the last step indicate that strains localize at the upper-left and the lower-right corners, as shown in Fig. 3.35. No convergence is possible after $\varepsilon_a = 0.75\%$.

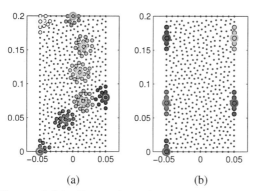

Figure 3.34: Examples of supports used for the evaluation of (a) spatial derivatives and (b) normal vectors on the pressure boundary.

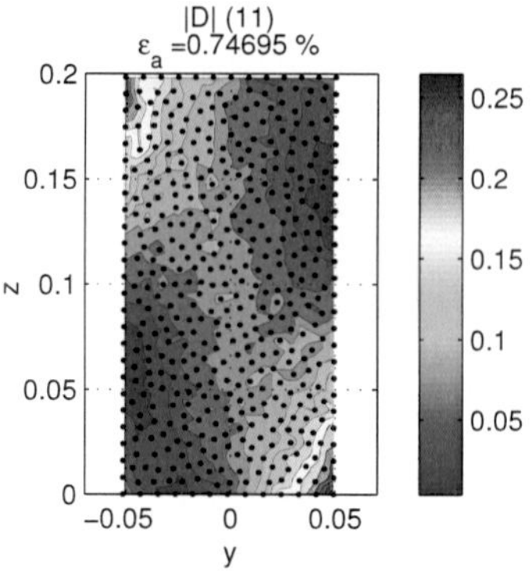

Figure 3.35: Contours of the $|\mathbf{D}|$ field at axial strain $\varepsilon_a = 0.75\%$.

The stress-strain curves for all particles are shown in Fig. 3.36. A homogeneous stress field prevails until $\varepsilon_a = 0.4\%$.

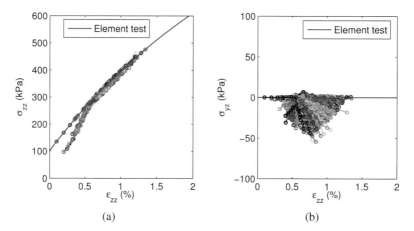

Figure 3.36: Stress-strain curves of all particles using solutions determined with the arc-length method. (a) σ_{zz} vs. ε_{zz} curves. (b) σ_{yz} vs. ε_{zz} curves.

3.4.3   First order polynomials with k-nn search

The support with $k = 6$ neighbors is used for the evaluation of the spatial derivatives and $k = 3$ for the evaluation of the normal vectors on the pressure boundary, as shown in Fig. 3.37. The solver only finds the solution of the first simulation step. The determined velocity field is shown in Fig. 3.38.

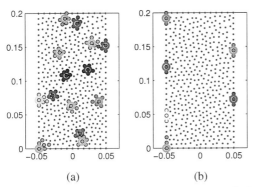

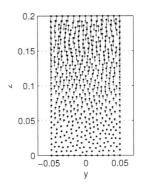

(a) (b)

Figure 3.37: Example of supports used for the evaluation of (a) spatial derivatives and (b) normal vectors on the pressure boundary.

Figure 3.38: Velocity field determined in the first step

3.4.4 Second order polynomials with k-nn search

The support with $k = 9$ neighbors is used for the evaluation of the spatial derivatives and $k = 5$ neighbors for the evaluation of the normal vectors on the pressure boundary, as shown in Fig. 3.39. The contours of the deformation field in the last step indicate that strains localize around the upper-right and lower-left corners as well as in the middle area of the bottom, as shown in Fig. 3.40.

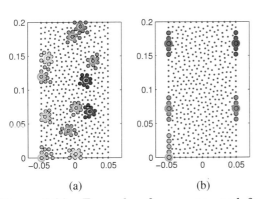

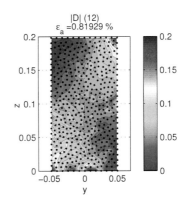

(a) (b)

Figure 3.39: Example of supports used for the evaluation of (a) spatial derivatives and (b) normal vectors on the pressure boundary.

Figure 3.40: Contours of the $|\mathbf{D}|$ field at axial strain $\varepsilon_a = 0.82\%$.

The stress-strain curves for all particles are shown in Fig. 3.41. A homogeneous stress field prevails for $\varepsilon_a < 0.4\%$

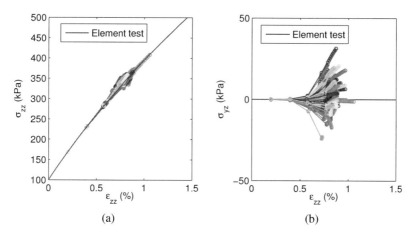

(a)   (b)

Figure 3.41: Stress-strain curves of all particles using solutions determined with the arc-length method. (a) σ_{zz} vs. ε_{zz} curves. (b) σ_{yz} vs. ε_{zz} curves.

3.4.5   Brief summary

With irregularly allocated particles, strain localization can be simulated but no shear bands appear. Inhomogeneous deformations occur at smaller axial strains compared with the results using regularly allocated particles. In addition, when strains localize, contours of the deformation field reveal that the patterns are more complex than the patterns in irregularly allocated particles. Using irregularly allocated particles is one of the basic ideas of the meshless method. However, in the current version of SPARC, it performs much worse than using a regular array of particles.

Chapter 4

Simulation of Conventional Laboratory Tests

This chapter is devoted to the demonstration of the SPARC framework using two conventional laboratory tests, the oedometer test and the triaxial test. In laboratory tests, the boundary conditions can be controlled and the results can be used for validation.

The simulations in this chapter use first-order polynomials with regularly allocated particles and small neighbor sizes determined by the fixed-radius search method. It has been shown in the parameter study (Chapter 3) that such combination performs the best in the biaxial test with the homogenous deformations being well simulated.

4.1   Oedometer test

Oedometer tests are known as element tests (with the friction between the oedometric ring and the specimen being neglected, see Fig. 4.1) in which the deformation in the specimen is homogeneous. In other words, a small part of the specimen of

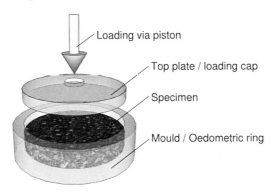

Figure 4.1: Loading system of the oedometer test. The mould prevents lateral deformation of the specimen.

any size is representative of the whole specimen. In such case, the form and the

size of the specimen do not affect the simulation results. The deformation in the oedometer test samples can be considered as a one-dimensional compression, as lateral
deformation is prohibited due to the confinement. In laboratory, an oedometer test
is conducted in a cylindrical mould and compressed (load-controlled) from the top
plate (Fig. 4.1). For simplicity, the cylindrical form of the specimen is replaced by a
rectangular box (consisting of 6 flat faces) in the simulation and the compression is
displacement-controlled.

4.1.1 Discretization of the specimen

The problem domain is discretized into soft particles. Totally, 64 ($4 \times 4 \times 4$) material
points are generated. The dimension of the specimen is $0.075 \times 0.075 \times 0.075$ m^3,
as illustrated in Fig. 4.2. With discretization, the position vector $\mathbf{x} = \begin{bmatrix} x_1 & x_2 & x_3 \end{bmatrix}$ for
each particle is obtained.

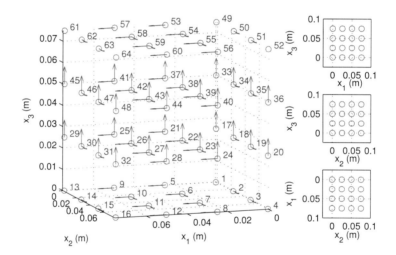

Figure 4.2: Discretization of the oedometer specimen into particles. The arrows
indicate the degrees of freedom. The numbers next to the particles are the particle
indices.

4.1.2 Initial values

Each particle carries material information, i.e. density (ρ, scalar), void ratio (e, scalar),
Cauchy stress tensor ($\mathbf{T}$, second order tensor), velocity ($\mathbf{v}$, vector) and position ($\mathbf{x}$,
vector). The initial values of ρ and e for all particles are prescribed as 1.6×10^3 kg/m^3
and 0.8, respectively.

4.1.2.1 Velocity field

Each particle has a velocity, $\mathbf{v}$:

$$\mathbf{v} = \begin{bmatrix} v_1 & v_2 & v_3 \end{bmatrix}, \tag{4.1}$$

where component 1 and component 2 are horizontal and component 3 is vertical. The particles on the top of the specimen, i.e. particle indices $49\ldots64$ (Fig. 4.2), are numerically marked as the top-plate particles which move constantly downward in the simulation. Thus, their velocities are prescribed as

$$\mathbf{v}_{\text{Top-plate particles}} = \begin{bmatrix} v_1^* & v_2^* & v_t \end{bmatrix} \text{m/s}, \tag{4.2}$$

where the '$*$' denotes *unknown* velocity components that need to be determined by solving the system of equations consisting of equilibrium equations. It is reasonable to prescribe the initial guessed values for the unknowns with $v_1^* = v_2^* = 0$ for all particles in the problem domain since the oedometer test is a one-dimensional compression test. The speed of the top plate is $v_t = -10^{-2}$ m/s in this simulation. Since barodesy is a rate-independent constitutive model, the plate velocity does not have an influence on the simulation results.

All particles on the bottom of the specimen (i.e. particle indices $1\ldots16$ in Fig. 4.2), are numerically marked as the bottom-plate particles. Since movements along the x_3-direction are prohibited, the third component of the velocity (v_3) of the bottom-plate particles is set to zero. The velocities at those particles are thus prescribed as

$$\mathbf{v}_{\text{Bottom-plate particles}} = \begin{bmatrix} v_1^* & v_2^* & 0 \end{bmatrix}. \tag{4.3}$$

For particles located on the boundary surface $x_1 = 0$, (i.e. particle indices $1\ldots4$, $17\ldots20$, $33\ldots36$, and $49\ldots52$) and on the surface $x_1 = 0.075$ (i.e., particle indices $13\ldots16$, $29\ldots32$, $45\ldots48$, and $61\ldots64$), their movements along x_2-axis are prohibited:

$$\mathbf{v} = \begin{bmatrix} v_1^* & 0 & v_3^* \end{bmatrix} \tag{4.4}$$

Likewise, movements of particles, located on the surface $x_2 = 0$ and on the surface $x_2 = 0.075$, along x_1-axis are prohibited:

$$\mathbf{v} = \begin{bmatrix} 0 & v_2^* & v_3^* \end{bmatrix} \tag{4.5}$$

Particles that are not on the boundaries have unknown velocity components in all directions:

$$\mathbf{v} = \begin{bmatrix} v_1^* & v_2^* & v_3^* \end{bmatrix} \tag{4.6}$$

The degrees of freedom of the system are shown in Fig. 4.2, indicated by arrows.

For all particle in the problem domain, the initial guessed values for v_3^* are prescribed as

$$v_3^* = \frac{x_3}{h_s} v_t \tag{4.7}$$

where h_s is the specimen height (Fig. 4.3).

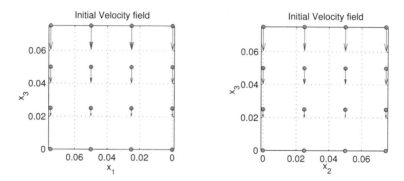

Figure 4.3: The initial guess of velocity field in views of x_1-x_3 and x_2-x_3 plane.

4.1.2.2   Stress field

In this simulation, the gravitational force is neglected as its influence is small when compared to the stress level the specimen undergoes during compression. Consequently, the initial stress tensor for all particles can be prescribed as:

$$\mathbf{T} = \begin{bmatrix} -5K_0 & 0 & 0 \\ 0 & -5K_0 & 0 \\ 0 & 0 & -5 \end{bmatrix} \text{(kPa)}, \tag{4.8}$$

where $K_0 = 0.5$ is taken as the coefficient of earth pressure at rest.

4.1.3   Numerical solver

The degrees of freedom of the system are 96 (Fig. 4.2). The unknowns are determined by solving the 96 equilibrium equations provided by eq. (2.2) using Newton's method (Section 2.7). Note that the gravitational force is neglected in this simulation and hence $\mathbf{g} = \mathbf{0}$ is applied in eq. (2.2). The Newton solver can solve the system of equations without convergence difficulties.

4.1.4 Time integration

Once the unknowns (velocity components) are determined, the stress tensors, positions, void ratios, and densities are advanced using the evolution equations eq. (2.9), eq. (2.10), eq. (2.11), and eq. (2.12), respectively. Note that $\mathbf{T}^{t+\Delta t}$ and $\mathbf{D}$ have been obtained in the solver once the velocity is determined, they may be directly used in these evolution equations.

4.1.5 Simulation results

The stress strain curves obtained using various number of particles and plate velocity are shown in Fig. 4.4. With boundary conditions that yield homogeneous deformation, the compression speed and the size of the problem or the number of particles used have no influences on the results. Consequently, the four curves in Fig. 4.4 overlap with one another.

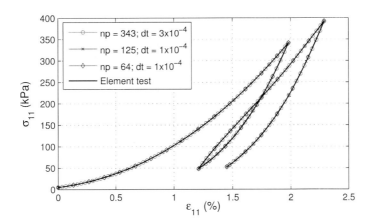

Figure 4.4: Stress-strain curves of various step sizes and various numbers of particles. Note that the curves are overlapping.

4.2 Triaxial test

In this section, the simulations of conventional triaxial CD tests using SPARC are detailed. Triaxial tests are widely used for the determination of mechanical parameters of geomaterials. Fig. 4.5 illustrates the shape of the specimen undergoing homogeneous and inhomogeneous deformation. The homogeneous deformation results from the frictionless boundaries and the homogeneous initial stress field and void ratio field. When there is friction between the plates and the specimen, the specimen forms a barrel shape and localized shear zones may appear in the specimen.

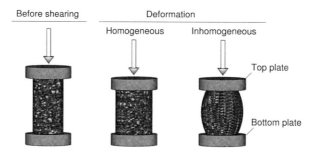

Figure 4.5: Illustration of triaxial tests under boundary conditions which yield homogeneous and inhomogeneous deformation.

4.2.1   Discretization of the specimen

The assumed dimensions of the cylindrical sample are radius = 5 cm and height = 20 cm. Totally, 444 material particles are generated with the configuration shown in Fig. 4.6. x_1 and x_2 are radial directions and x_3 is the axial direction.

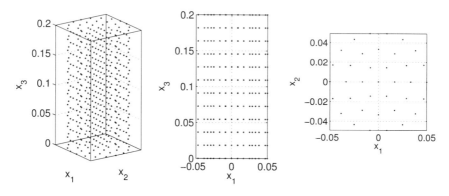

Figure 4.6: Discretization of a triaxial test specimen in the form of a cylinder.

4.2.2   Boundary conditions and degrees of freedom

The boundary conditions determine whether the deformation of the sample can be homogeneous. When friction exists between the soil and the plates, inhomogeneous deformation is expected. Assuming friction, top plate particles (Fig. 4.7a) are fixed in horizontal directions with their velocities prescribed as follows:

$$\mathbf{v} = \begin{bmatrix} 0 & 0 & v_t \end{bmatrix} \tag{4.9}$$

where v_t is a prescribed top plate velocity. For particles on the bottom plate:

$$\mathbf{v} = \begin{bmatrix} 0 & 0 & 0 \end{bmatrix} \tag{4.10}$$

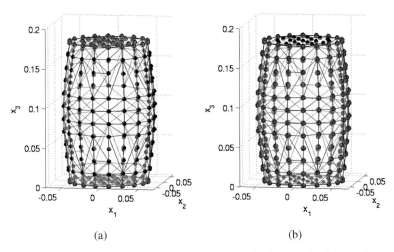

(a)   (b)

Figure 4.7: Illustrations of (a) plate particles (marked as red) (b) surface particles (marked as blue).

When the plates are frictionless, the velocities of particles on the top plate for this condition are prescribed as follows:

$$\mathbf{v} = \begin{bmatrix} v_1^* & v_2^* & v_t \end{bmatrix} \tag{4.11}$$

where v_t is the prescribed top plate velocity and '*' denotes unknown variables. For particles on the bottom plate:

$$\mathbf{v} = \begin{bmatrix} v_1^* & v_2^* & 0 \end{bmatrix} \tag{4.12}$$

For frictionless plates, the particle in the middle of the top plate is also fixed to prevent horizontal rigid body translation of the sample. In addition, the plate particles on the plane $x_2 = 0$ are prescribed with $v_1 = 0$ to prevent rigid body rotation of the sample.

The particles not on the plates are free to move in all directions:

$$\mathbf{v} = \begin{bmatrix} v_1^* & v_2^* & v_3^* \end{bmatrix} \tag{4.13}$$

The degrees of freedom for simulations with frictionless plates and with fixed plate particles are shown in Fig. 4.8, with the line segments indicating the degrees of freedom.

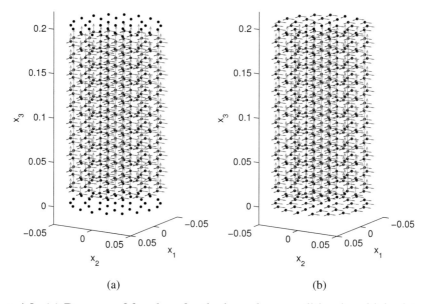

(a)   (b)

Figure 4.8: (a) Degrees of freedom for the boundary condition in which plate particles are fixed. (b) Degrees of freedom for frictionless-plate conditions.

4.2.3 Governing equations

Constant cell pressure is applied to the particles between the cell water and the specimen. These particles are called surface particles (Fig. 4.7b). The movements of these particles are governed by eq. (2.122). The remaining particles are governed by eq. (2.2) with $\mathbf{g} = \mathbf{0}$. The simulation results from solutions determined by the arc-length method are demonstrated herein.

4.2.4 Initial conditions

The initial condition of the simulation is listed in Table 4.1.

Table 4.1: Initial condition of the simulation.

Material	Initial density ρ_0 (kg/m^3)	Initial void ratio, e_0 (-)	Cell pressure, p (Pa)
Dense sand	$1.6234 \cdot 10^3$	0.6324	$-100 \cdot 10^3$

4.2.5 Simulation results

4.2.5.1 Convergence

For the simulation with frictionless plates, convergence at axial strain $\varepsilon_a \geq 5.6\%$ is not possible using the arc-length method. With Newton's method and Levenberg-Marquardt method, convergence is not possible even before the peak of the stress-strain curve ($\sim 4\%$).

For the simulation with fixed plate particles, the sample is can be compressed up to $\varepsilon_a = 13\%$ Thereafter, convergence is not possible even with a very small time increment $\Delta = 10^{-7}$s.

4.2.5.2 Stress-strain curve

The stress-strain curves resulting from homogeneous and inhomogeneous deformation are shown in Fig. 4.9. The curve of the simulation with homogeneous deformation overlaps with the element test until $\varepsilon_a = 4.7\%$. Thereafter, due to the accumulated errors in the solutions, strains start to localize in the middle of the sample, as will be shown in the next sub-section. It is noted that the number of neighbors in particles' supports do not change in this simulation.

For the simulation with inhomogeneous deformation, σ_a of the stress-strain curve is obtained using the averaged value of σ_z of the plate particles excluding those on the edges of the top plate, as shown in Fig. 4.10. These edge particles are excluded due to stress concentration on the plate edges, as will be shown in Section 4.2.5.4. σ_z is the normal stress at the plate particles:

$$\sigma_z = \mathbf{t}\mathbf{n} \tag{4.14}$$

$$= (\mathbf{T}\mathbf{n})\mathbf{n} \tag{4.15}$$

where $\mathbf{n} = [0\ 0\ 1]^\mathrm{T}$ is the normal vector of the top plate surface and $\mathbf{t}$ is the traction on the top plate surface. The curves in the simulations with inhomogeneous deformation show discrepancies when compared with the element test.

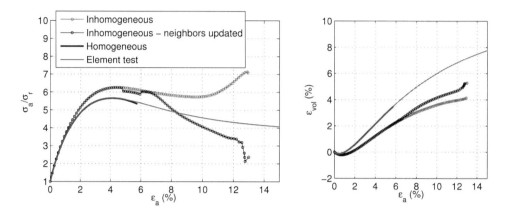

Figure 4.9: (a) Stress-strain curves. (b) Volumetric strain-axial strain curve. The pseudo-arc-length method is used as the numerical solver.

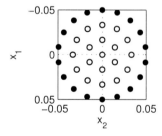

Figure 4.10: Illustration of particles (hollow dots) used for the evaluation of σ_a for the inhomogeneous deformed sample.

The particles move during the simulation and some particles therefore change their neighbors. The total number of particles that have their neighbors changed during the simulation is listed in Table 4.2. 48 particles have different numbers of neighbors

Table 4.2: Total number of particles whose neighbors change during the simulation.

Total Number of particles	ε_a at which neighbors changed
48	4.84%
24	6.07%
24	6.49%
2	12.99%

at $\varepsilon_a = 4.84\%$, 24 particles have different numbers of neighbors at $\varepsilon_a = 6.07\%$ and $\varepsilon_a = 6.49\%$, and 2 particles have different number of neighbors at $\varepsilon_a = 12.99\%$. The curve 'Inhomogeneous - neighbors updated' in 'Fig. 4.9 shows that, when the neighbors in supports are not the same as those in the previous time increment, the stress-strain curve exhibits either a jump (at $\varepsilon_a = 4.84\%$, 6.07%, and 12.99%) or significant softening (at $\varepsilon_a = 6.49\%$). This reveals that, when the number of neighbors in supports change in the current simulation step, it has a relevant influence to the system. The same simulation without updating the neighbors has been carried our for comparison. When the neighbors are not updated in all simulation steps, the curve is smooth (i.e. the gray line with solid dots in Fig. 4.9).

4.2.5.3   Deformation field

Fig. 4.11 illustrates the deformation fields (visualized by principal directions of $\mathbf{D}$) in the simulation with frictionless plates at various axial strains, using particles on a vertical slice in the middle of the specimen, on the plane $x_2 = 0$. In this simulation, the homogeneity in deformation holds until the stress-strain curve reaches the peak, thereafter deformation begins to localize (Fig. 4.11d).

The deformation fields in the simulation with plate particles being fixed are shown in Fig. 4.12. With time, the deformation concentrates in the middle of the specimen, whereas the zones near the middle of the plates have little deformation. The contour plot of $|\mathbf{D}|$ (Fig. 4.13) reveals that two rigid cones are formed. Significantly larger deformation occurs on the edges of the plates. During the compression, the specimen obtains a barrel shape, as shown in Fig. 4.12h and in Fig. 4.14. The principal directions of $\mathbf{D}$ rotate (Fig. 4.12h), with one of the directions perpendicular to the specimen surface.

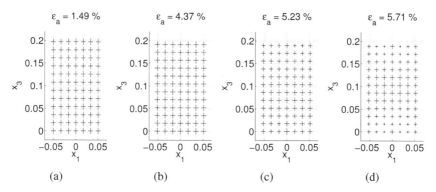

Figure 4.11: Homogeneous deformation fields of the simulation with frictionless plates. Particles on the plane $x_2 = 0$ are used in the plots.

Fig. 4.15 shows the evolution of particle movements in the simulation with fixed plate particles at various strains using the norm of the velocity.

$$|\mathbf{v}| = \sqrt{v_1^2 + v_2^2 + v_3^2} \qquad (4.16)$$

These figures reveal that the particles at the plates move with the same speed as the plates. With time, two rigid conical blocks are formed. These cones are captured by the X-ray tomography on triaxial test samples.

4.2.5.4 Stress field

For the simulation with frictionless plates, the stress field is homogeneous, as shown in Fig. 4.16. The directions begin to rotate slightly after the peak of the stress-strain has been reached in this simulation.

For the simulation with plate particles being fixed on the plates, the stress significantly concentrated on the edges of the plates during compression, as shown in Fig. 4.17. The rotation of the principal directions can be seen by plotting only the directions without the magnitudes of the principal stresses, as shown in Fig. 4.18. Due to the fact that the plate particles are fixed, the principal directions of stress tensor rotate when the simulation begins. The principal directions on the surface particles are perpendicular to the specimen surface due to the cell pressure.

4.2.5.5 Void ratio

The evolution of void ratio in the simulation with fixed plate particles is shown in Fig. 4.19. The initial void ratio is 0.6324 (Table 4.1). The void ratio increases in

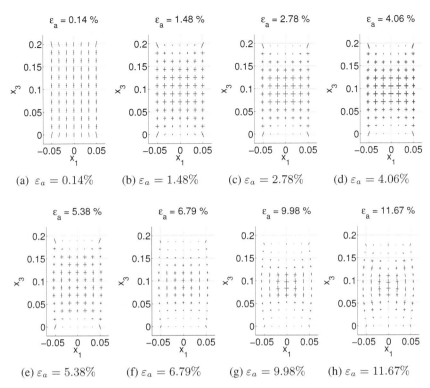

(a) $\varepsilon_a = 0.14\%$ (b) $\varepsilon_a = 1.48\%$ (c) $\varepsilon_a = 2.78\%$ (d) $\varepsilon_a = 4.06\%$

(e) $\varepsilon_a = 5.38\%$ (f) $\varepsilon_a = 6.79\%$ (g) $\varepsilon_a = 9.98\%$ (h) $\varepsilon_a = 11.67\%$

Figure 4.12: Principal directions of $\mathbf{D}$ in the simulation with fixed plate particles are plotted using particles on the plane $x_2 = 0$. The lengths of the crosses indicate the magnitudes of the principal strain rates.

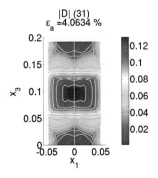

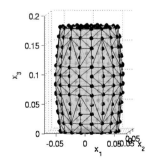

Figure 4.13: Contours of $|\mathbf{D}|$. Particles on the plane $x_2 = 0$ are used in the plots.

Figure 4.14: Barrel-shaped triaxial sample at $\varepsilon_a = 11.8\%$. The hull of the specimen is illustrated using the plate particles and surface particles.

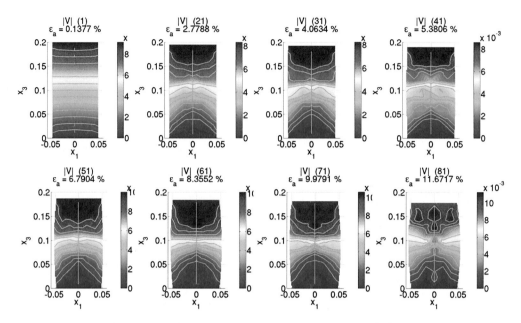

Figure 4.15: Evolution of velocity in terms of the contours of $|\mathbf{v}|$ at various strains (Table 4.1, dense sand).

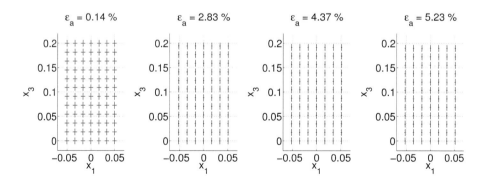

Figure 4.16: Evolution of principal directions of the stress tensors in the simulation with frictionless plates. The lengths of the crosses indicate the relative magnitudes of the principal stresses.

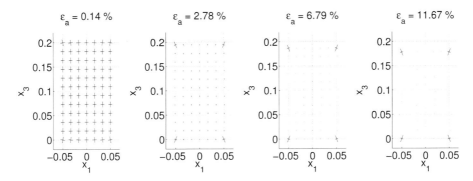

Figure 4.17: Evolution of principal directions of the stress tensors in the simulation with fixed plate particles at selected axial strains. The lengths of the crosses indicate the relative magnitudes of the principal stresses. Stresses significantly concentrate on the edge of the plates due to the fixed plate particles.

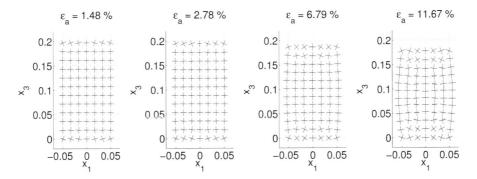

Figure 4.18: Evolution of principal directions of the stress tensors at selected axial strains. The magnitudes of the principal stresses are *not* illustrated in order to show the rotation of the principal directions.

most parts of the sample, except from the edges of the plates. In the middle area of the specimen, the void ratios significantly increase up to 0.8. On the plates, two conical blocks with $e \approx 0.635$ appear. In these two rigid zones, the particles have little relative movements with respect to the plates (Fig. 4.15) and little deformation occurs zones (Fig. 4.12). Consequently, the void ratio $e \approx 0.635$ in these two conical zones has little change.

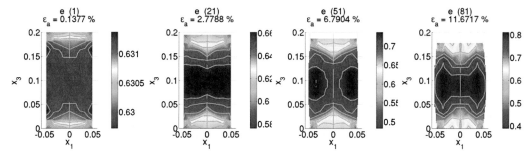

Figure 4.19: Contours of void ratio (e) on the plane $x_2 = 0$.

Chapter 5

Validation

The validity of a simulation mainly relies on (1) the underlying constitutive model and (2) the numerical method that evaluates the spatial derivatives and solves the governing equations of a problem subject to the applied boundary conditions. Strictly speaking, there is no correct or wrong model as each one more or less compromises the complex nature by simplifying its principles. The validation of SPARC is carried out by confirming the rationality of the simulation results. Note that the validation of the constitutive model is not in the scope of SPARC. For instance, (1) the stress-strain curve obtained using SPARC are compared with the underlying constitutive model (element test). (2) Principal directions of stresses and incremental strains are examined. (3) The deformation field is compared with the X-ray tomography results. (4) Results from FEM and SPARC are compared.

5.1   Oedometer test

When the boundary conditions are prescribed in a way such that the deformation fields are homogeneous, the simulation results shall be comparable with those obtained from the integration of the underlying constitutive model. In the simulations of oedometer tests, the boundaries are frictionless, resulting in the ideal element test condition. Fig. 5.1 shows that the stress-strain curves of the SPARC simulations and the barodesy element test are in good agreement.

Oedometer tests (section 4.1) are carried out by confining lateral displacements (discarding frictions between the specimen and the mould (Fig. 4.1)) and, therefore, it is a one-dimensional deformation (along z-axis), implying that the principal directions of incremental strains shall not rotate during the simulation. Consequently, as shown in Fig. 5.2, one direction of principal strain directions is oriented parallel to z-axis and the other two directions of principal strain are perpendicular to z-axis. These figures also reveal that the strain field is homogeneous during compression.

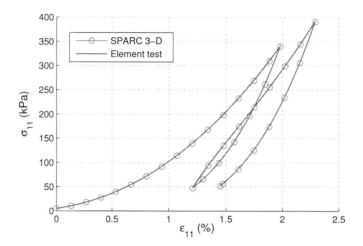

Figure 5.1: Comparisons between the stress-strain curves obtained from SPARC 3D simulation (demonstrated in section 4.1) and the element test simulation using barodesy.

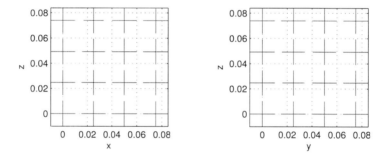

Figure 5.2: Principal directions of incremental strain ($\mathbf{D}\Delta t$) in the views of x-z and y-z. The displacements are confined in x- and y-direction and loaded in z-direction.

5.2 Triaxial test

5.2.1 Homogeneous deformation

The results of the triaxial test simulation with frictionless plates (boundary conditions detailed in Section 4.2) are used. As a result of the boundary conditions, one of the directions shall be parallel to x_3-axis, and the principal strain directions shall not rotate in the x_1-x_3 plane, as shown in Fig. 5.3.

The stress strain curve of all particles shall be identical and overlap with the element test simulation results before inhomogeneous deformation begins, as shown in Fig. 5.4.

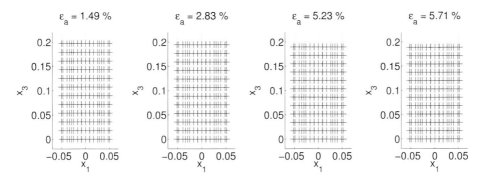

Figure 5.3: Evolution of principal directions of **D**. The length of the crosses indicate the magnitude of the principal strains.

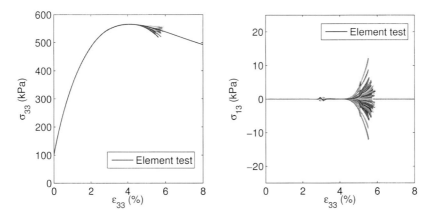

Figure 5.4: Stress-strain curves of all particles. Stress components σ_{33} and σ_{13} vs. strain component ε_{33}.

5.2.2   Inhomogeneous deformation

The incremental principal directions obtained in the simulation with fixed plate particles are qualitatively in good agreement with the results obtained using the X-ray tomography on a conventional triaxial test carried out by Kolymbas [27]. In that experiment, due to the friction between the plates and the material, conical zones with significantly less deformation appeared.

5.3   Simulation with FE method

A finite-element (FE) simulation of a biaxial test using barodesy for clay [46] was carried out in [12]. Barodesy for clay are implemented in ABAQUS through 'umat'.

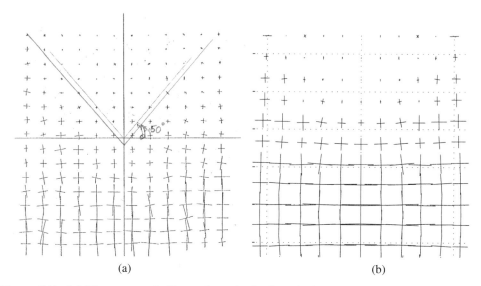

                              (a)   (b)

Figure 5.5: (a) The crosses indicate the principal strain increments during shear [27].
(b) Incremental principal strain at $\varepsilon_a = 4.4 \rightarrow 4.8\%$ using SPARC in a simulation
with the specimen discretized into 1710 particles and cell pressure $\sigma_c = 100$ kPa.

10×20 four-node elements are used in the simulation discretizing the sample with
dimensions 10 cm $\times$ 20 cm (width and height).[1] The simulation was carried out
under plane strain condition using geometrical nonlinear analysis.[2] The boundary
conditions are as follows (see Fig. 3.1b): Pressure is applied to the free surface. The
boundaries on the top and bottom of the sample are frictionless. One node in the
middle of the top of the sample is fixed horizontally to prevent horizontal rigid body
translation of the sample.

The obtained stress-strain curves at all nodes are shown in Fig. 5.6a. The component
σ_{zz} in the stress tensor $\mathbf{T}$ and the component ε_{zz} in the strain tensor ε are used. The
curves overlap with one another before the peaks ($\varepsilon_{zz} = 1.6\%$) are reached, indi-
cating that the stress field is homogeneous until $\varepsilon_{zz} = 1.6\%$. These curves are also
in good agreement with the curve of element test which is plotted for comparison.
However, the stresses σ_{zz} around the peaks are slightly larger than the curve of element test. After the peaks, the stress-strain curves diverge. The simulation aborts at
axial strain $\varepsilon_a = 4.4\%$ when the prescribed minimum time increment $\Delta t = 10^{-10}$s
is reached. During the simulation, for $\varepsilon_a < 1.6\%$, the time increment is $\Delta t = 0.01$s.
From $\varepsilon_a = 1.6\%$ to 3.1%, Δt is decreased to 0.001s. Thereafter from $\varepsilon_a = 3.1\%$ to

---

[1]The implementation of barodesy for clay in the Umat and the FE simulation were conducted by
our colleague Dr. Barbara Schneider-Muntau, University of Innsbruck.
   [2]In the geometrical nonlinear analysis, the strain tensor of the element is evaluated in the deformed
configuration.

4.4%, Δt is increased to 0.01s. At $\varepsilon_a = 4.4\%$, Δt is decreased down to 10^{-10}s and convergence is not possible.

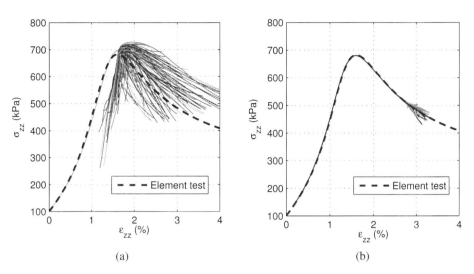

(a)   (b)

Figure 5.6:   Stress-strain curves at all nodes or particles obtained using (a) FE method and (b) SPARC.

A simulation with the same boundary conditions and the same constitutive equation was carried out using SPARC for comparison. The particle configuration used in SPARC corresponds to the configuration of the nodes used in the FE simulation. The neighbors and the polynomials used for the evaluation of the spatial derivatives are the same as the benchmark described in Chapter 3. The stress-strain curves of all particles are shown in Fig. 5.6b. This figure shows that the stress field is homogeneous for $\varepsilon_{zz} < 2.8\%$, and are in good agreement with the curve of the element test. At $\varepsilon_{zz} \geq 2.8\%$, the stress field becomes inhomogeneous and the step size decreases from $4 \cdot 10^{-2}$s to $4 \cdot 10^{-11}$s in the following 14 steps thereafter. For $\varepsilon_a \geq 3.0\%$, convergence is not possible even with a very small step size.[3]

The curves obtained using FE and SPARC (i.e. Fig. 5.6a and b respectively) before the onset of inhomogeneous deformation are in good agreement. Inhomogeneous stress field appears when the sample is compressed to some extend. SPARC predicts preciser the stresses σ_{zz} around the peaks when compared with the stress-strain curve of an element test. After the peaks, the curves obtained by FE method diverge; and the curves obtained by SPARC overlap with one another up to $\varepsilon_a = 2.8\%$.

---

[3]When the deformation field is homogeneous, the component ε_{zz} in the strain tensor is the same as the axial strain ε_a. When the deformation field is inhomogeneous, the values of ε_{zz} are different at particles. In this case, ε_a is used for description.

Chapter 6

Simulation of the 'Zig-Zag'-Tests

Model tests using the zig-zag apparatus were conducted. The details of the experiment and the results are documented in Appendix J. The experiment involves large deformation. Simulations of this experiment are carried out. The boundary conditions and the simulation results are detailed in this chapter.

6.1 Boundary conditions

The boundary conditions are illustrated in Fig. 6.1.

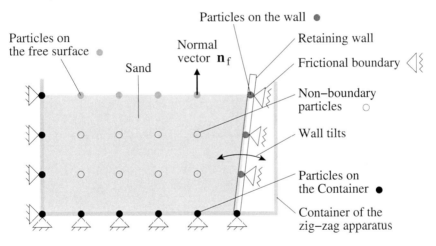

Figure 6.1: Boundary conditions for the model test with the zig-zag apparatus.

For the particles on the free surface, two governing equations are tested. The one requires that the traction vanishes at the free boundary.

$$\mathbf{T}^{t+\Delta t}\mathbf{n}_{\mathrm{f}}^{t+\Delta t} = \mathbf{0} \tag{6.1}$$

where $\mathbf{n}_{\mathrm{f}}^{t+\Delta t}$ is the unit vector normal to the free surface. Under plane-strain condition, this vector equation provides two scalar equations to determine the two unknown velocity components of every surface particle. An alternative to the equation above is to require that the stress tensor vanished at surface particles

$$\mathbf{T}^{t+\Delta t} = \mathbf{0} \qquad (6.2)$$

It results from the fact that, for cohesionless materials, all components of stress at a free surface vanish. Under plane-strain condition, this equation provides four scalar equations, namely $T_{11}^{t+\Delta t} = 0$, $T_{22}^{t+\Delta t} = 0$, $T_{12}^{t+\Delta t} = 0$, and $T_{33}^{t+\Delta t} = 0$ (assuming directions 1 and 2 define the 2D modeling space).

For non-boundary particles (marked as hollow dots), the governing equation reads $\nabla^{t+\Delta t} \cdot \mathbf{T}^{t+\Delta t} + \rho^{t+\Delta t}\mathbf{g} = \mathbf{0}$. Particles marked as black have known velocities. For example, those on the frame are fixed to the container and hence their velocity is prescribed as zero.

Particles on the tilting retaining wall are subject to frictional boundary condition (dry friction), as shown in Fig. 6.2. We consider a particle on the retaining wall, particle i_{w}. At time t, the position of the wall is inclined by θ^t and the wall tilts

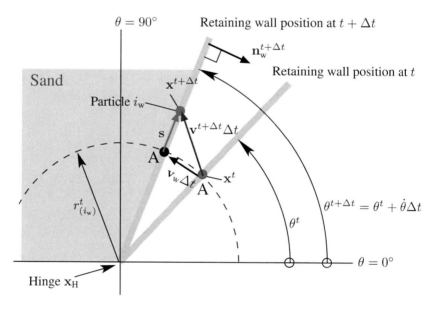

Figure 6.2: Frictional boundary condition on the tilting wall

at an angular speed $\dot{\theta}$. In the time interval from t to $t + \Delta t$, the particle i_{w} moves from position $\mathbf{x}^t$ (point A) to $\mathbf{x}^{t+\Delta t}$. The distance between point A and the hinge is $r_{(i_{\mathrm{w}})}^t = |\mathbf{x}^t - \mathbf{x}_{\mathrm{H}}|$ at all times. Thus, given a time increment Δt, the current wall position θ^t, the angular speed of tilts $\dot\theta$, and $r^t_{(i_w)}$, point A of the wall moves from time t to $t + \Delta t$ with the velocity (see Appendix L for derivation):

$$\mathbf{v_w} = \frac{r_{(i_w)}}{\Delta t} \left[\cos(\theta^t + \dot\theta\Delta t) - \cos\theta^t, \quad \sin(\theta^t + \dot\theta\Delta t) - \sin\theta^t \right] \tag{6.3}$$

Consequently, particle i_w moves from $\mathbf{x}^t$ to $\mathbf{x}^{t+\Delta t}$

$$\mathbf{x}^{t+\Delta t} = \mathbf{x}^t + \mathbf{v_w}\Delta t + \mathbf{s} \tag{6.4}$$

$\mathbf{s}$ is the slip along the wall

$$\mathbf{s} = \left(\mathbf{v}^{t+\Delta t} - \mathbf{v_w} \right) \Delta t \tag{6.5}$$

The frictional force acting on the particle i_w reads

$$\mathbf{F} = -\mathbf{s}^o |\mathbf{N}| \mu \left(1 - \exp\left(-\alpha\frac{|\mathbf{s}|}{s_0} \right) \right) \tag{6.6}$$

where $\mathbf{s}^o$ is the unit vector of $\mathbf{s}$ ($\mathbf{s}^o = \mathbf{s}/|\mathbf{s}|$); $|\mathbf{N}|$ is the normal force; μ is the coefficient of friction; α is a mobilization factor that defines how much the friction is mobilized; and s_0 is the slip distance required to mobilize a certain percentage of full friction. By assuming the exponential mobilization of eq. (6.6) and choosing $\alpha = 4.6052$, s_0 is the slip distance required to mobilize 99% of full friction ($|\mathbf{N}|\mu$), as shown in Fig. 6.3.[1] In this simulation, $s_0 = 1$ mm is used.

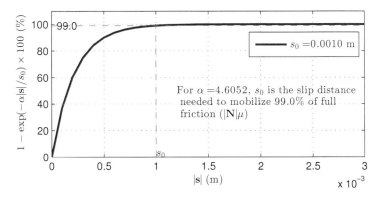

Figure 6.3: Mobilization $\left(1 - \exp\left(-\alpha\frac{|\mathbf{s}|}{s_0} \right) \right)$ as a function of the slip distance $|\mathbf{s}|$ of a particle on the frictional surface. With $\alpha = 4.6052$, s_0 is the slip mobilizing 99% of full friction.

---

[1] 99% of full friction means: For $|\mathbf{s}| = s_0$, $1 - \exp(-\alpha \cdot 1) = 0.99$. It yields $\alpha = 4.6052$.

μ is computed as:

$$\mu = \tan(\phi_\mathrm{w}) \tag{6.7}$$

where ϕ_w is the wall friction angle. $\phi_\mathrm{w} = \phi/3$ is assumed in the simulation because the wall surface is relatively smooth. The friction angle ϕ of the test sand is $34°$.

Given the unit vector normal to the contact surface at time $t + \Delta t$ (Fig. 6.2):

$$\mathbf{n}_\mathrm{w}^{t+\Delta t} = \left[\sin(\theta^{t+\Delta t}), \ -\cos(\theta^{t+\Delta t})\right] \tag{6.8}$$

the shear stress due to dry friction at the wall reads:

$$\boldsymbol{\tau} = \mathbf{t} - \boldsymbol{\sigma} = -\mathbf{s}^o|\boldsymbol{\sigma}|\mu\left(1 - \exp\left(-\alpha\frac{|\mathbf{s}|}{s_0}\right)\right)$$

$$\rightsquigarrow \quad \mathbf{t} - \boldsymbol{\sigma} + \mathbf{s}^o|\boldsymbol{\sigma}|\mu\left(1 - \exp\left(-\alpha\frac{|\mathbf{s}|}{s_0}\right)\right) = \mathbf{0} \tag{6.9}$$

where $\mathbf{t} = \mathbf{T}^{t+\Delta t}\mathbf{n}_\mathrm{w}^{t+\Delta t}$ is the surface traction on the contact surface; and $\boldsymbol{\sigma} = (\mathbf{t} \cdot \mathbf{n}_\mathrm{w}^{t+\Delta t})\mathbf{n}_\mathrm{w}^{t+\Delta t}$ is the normal stress on the contact surface. This vector equation provides two scalar equations. In addition, since particle i_w slips on the contact surface, the following condition must be fulfilled as well:

$$\mathbf{s} \cdot \mathbf{n}_\mathrm{w}^{t+\Delta t} = 0 \tag{6.10}$$

As a result, to determine $\mathbf{v}^{t+\Delta t}$ for particles on the frictional boundary, eqs. (6.9) and (6.10) must be satisfied simultaneously.

6.2   Simulation set-ups

Table 6.1 lists six possible set-ups of boundary conditions and the corresponding governing equations, particularly, for particles on the free surface and on the frictional wall.[2] The figures in Table 6.1 refer to the graph of Fig. 6.1. At surface particles (marked as light gray dots), the governing equation is either eq. (6.1) or eq. (6.2). At wall particles (marked as dark gray), the governing equation for the frictional boundary condition is eq. (6.9). All particles on the wall must satisfy eq. (6.10) if the velocities of these particles are unknown. Black dots in BC5 and BC6 denote that the velocities are given.

---

[2]The velocities for the particles on the container are zero and the governing equation for non-boundary particles is $\nabla^{t+\Delta t} \cdot \mathbf{T}^{t+\Delta t} = \mathbf{0}$.

Table 6.1: List of boundary conditions and the corresponding governing equations used for particles particularly on the free surface and the retaining wall. The figures in this table refer to the graph of Fig. 6.1.

Case	Governing equations
BC1	$\mathbf{T}^{t+\Delta t} = \mathbf{0}$ (eq. 6.2) $\mathbf{t} - \sigma + \mathbf{s}^o|\sigma|\mu\left(1 - \exp\left(-\alpha\frac{|\mathbf{s}|}{s_0}\right)\right) = \mathbf{0}$ (eq. 6.9) $\mathbf{s} \cdot \mathbf{n}_w^{t+\Delta t} = 0$ (eq. 6.10)
BC2	$\mathbf{T}^{t+\Delta t}\mathbf{n}_f^{t+\Delta t} = \mathbf{0}$ (eq. 6.1) $\mathbf{t} - \sigma + \mathbf{s}^o|\sigma|\mu\left(1 - \exp\left(-\alpha\frac{|\mathbf{s}|}{s_0}\right)\right) = \mathbf{0}$ (eq. 6.9) $\mathbf{s} \cdot \mathbf{n}_w^{t+\Delta t} = 0$ (eq. 6.10)
BC3	$\mathbf{T}^{t+\Delta t} = \mathbf{0}$ (eq. 6.2) $\mathbf{t} - \sigma + \mathbf{s}^o|\sigma|\mu\left(1 - \exp\left(-\alpha\frac{|\mathbf{s}|}{s_0}\right)\right) = \mathbf{0}$ (eq. 6.9) $\mathbf{s} \cdot \mathbf{n}_w^{t+\Delta t} = 0$ (eq. 6.10)
BC4	$\mathbf{T}^{t+\Delta t}\mathbf{n}_f^{t+\Delta t} = \mathbf{0}$ (eq. 6.1) $\mathbf{t} - \sigma + \mathbf{s}^o|\sigma|\mu\left(1 - \exp\left(-\alpha\frac{|\mathbf{s}|}{s_0}\right)\right) = \mathbf{0}$ (eq. 6.9) $\mathbf{s} \cdot \mathbf{n}_w^{t+\Delta t} = 0$ (eq. 6.10)
BC5	$\mathbf{T}^{t+\Delta t} = \mathbf{0}$ (eq. 6.2) Velocities are prescribed using eq. (6.3) (no slip along the wall)
	Continued on next page

Table 6.1 : (*continued*)

Case	Governing equations
BC6	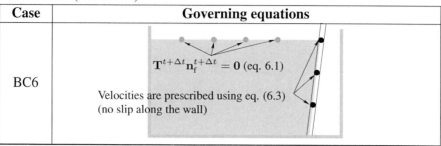 $\mathbf{T}^{t+\Delta t}\mathbf{n}_{\mathrm{f}}^{t+\Delta t} = 0$ (eq. 6.1) Velocities are prescribed using eq. (6.3) (no slip along the wall)

The particle configuration of the simulations (with various boundary conditions shown in Table 6.1) and neighbors obtained using the fixed-radius neighbor search with a radius of 0.0094 m are shown in Fig. 6.4. The simulation domain has the width $w_z = 0.2$ m and height $h_z = 0.1$ m.

First order polynomials (eq. 2.18) are used for the evaluation of spatial derivatives. The stop criterion $\epsilon \leq 10^{-2}$ is chosen for the numerical solver. The solver adopts pseudo-arc-length method, as described in Section 2.7.4.4. The initial guess of the velocity field is shown in Fig. 6.5, with an initial value of angular velocity $\dot{\theta} = 10°$ 1/sec (inward-tilts, pushing against sand) for wall tilts and an initial position of the wall $\theta = 90°$ being used. Based on the coordinate axes used in Fig. 6.5, the initial velocity field is obtained by evaluating the velocity components v_1 and v_2 of the particles as follows: For a particle,

$$\begin{cases} v_1 = \frac{x_1 + w_z/2}{w_z} \cdot \frac{x_1 + h_z}{h_z} \cdot v_{(\text{wall tip})} \\ v_2 = 0 \end{cases} \tag{6.11}$$

x_1 is a component of the position vector of the particle under evaluation; and $v_{(\text{wall tip})}$ (computed using eq. 6.3) is the velocity of the particle on the wall surface as well as on the free surface. The tilt amplitude of the wall in the simulation is $1°$ (i.e. $90° \pm 1°$).

The initial time increment is $\Delta t = 10^{-4}$ s. Based on the experimental results (Appendix J), the initial void ratio $e = 0.77$ and the initial density $\rho = 1.5$ g/cm^3 are used in the simulations. The resulting initial T_{22} ($= \rho g z$, with z being the depth from the free surface and $g = -9.81$ m/s^2) field is shown in Fig. 6.6.

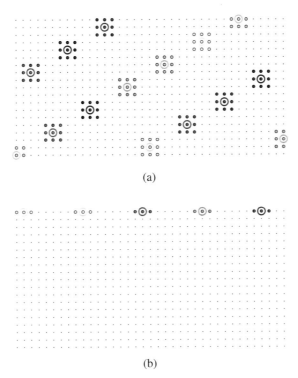

(a)

(b)

Figure 6.4: Examples of neighbors used for the evaluation of (a) $\mathbf{L}$ and $\nabla \cdot \mathbf{T}$, and (b) unit vector $\mathbf{n}_f$ normal to the free surface. Particles with the same color are neighbors of the circled one.

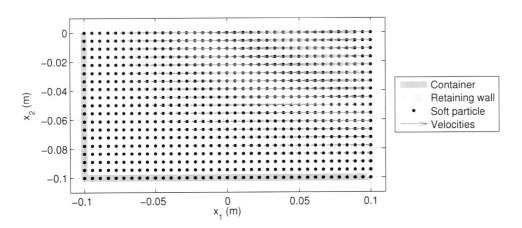

Figure 6.5: Initial guess of the velocity field for the first time increment.

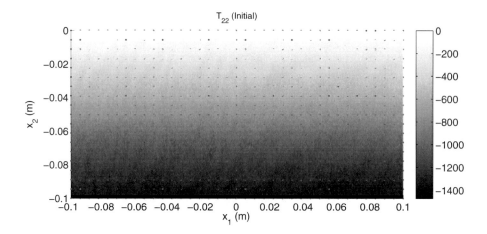

Figure 6.6: Initial T_{22} field. Unit in Pa.

6.3   Results

The obtained velocity fields of the simulations for the first time increment, with the boundary conditions summarized in Table 6.2 and simulation set-ups described in Section 6.2, are shown in Table 6.2. The velocity fields in cases BC1, BC2, BC3 and BC4 are extremely non-smooth, especially the particles near the free surface. On the contrary, in cases BC5 and BC6, the velocity fields are relatively smoother compared to cases BC1 through BC4, and both velocity fields are comparable with each other.

Table 6.2: Obtained velocity fields at the first time increment for the simulations with boundary conditions shown in Table 6.1 and simulation set-ups detailed in Section 6.2.

Case	Obtained solution (Velocity field)
BC1	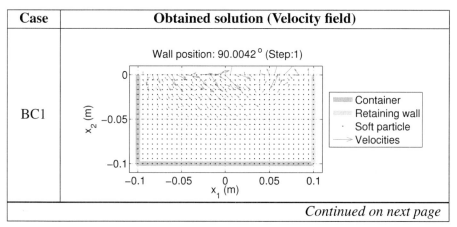
	Continued on next page

Table 6.2 : (*continued*)

Case	Obtained solution (Velocity field)
BC2	
BC3	
BC4	
BC5	

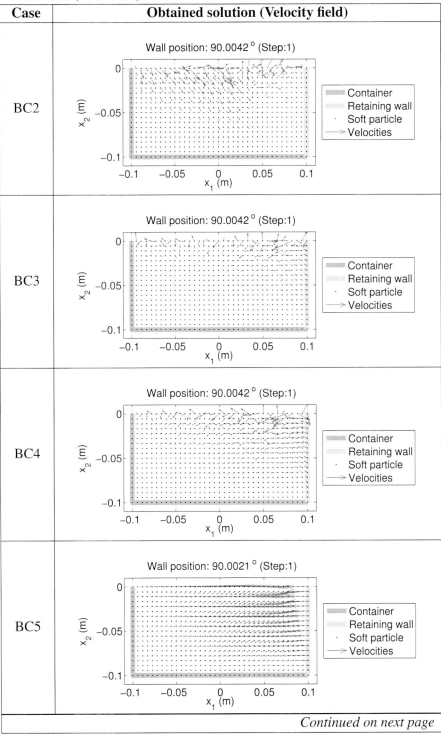

Continued on next page

Table 6.2 : (*continued*)

Case	Obtained solution (Velocity field)
BC6	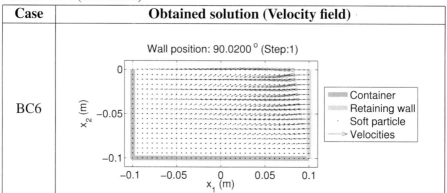

With increasing simulation steps, the velocity fields in the upper part of the simulation domain become even more chaotic for cases BC1 through BC4, as can be seen in Fig. 6.3. The simulations were manually aborted after $\Delta t < 10^{-6}$ s. The velocity fields in cases BC5 and BC6, on the contrary, are relatively smoother. Comparing results shown in Table 6.2 and Table 6.3, it can be seen that when the particles are fixed on the wall with their velocities prescribed using eq. (6.3), the obtained velocity field is smoother.

Table 6.3: Obtained velocity fields at a later simulation step for the simulations with boundary conditions shown in Table 6.1 and simulation set-ups detailed in Section 6.2. Note that the simulation of case BC4 did not last more than 50 steps.

Case	Obtained solution (Velocity field)
BC1	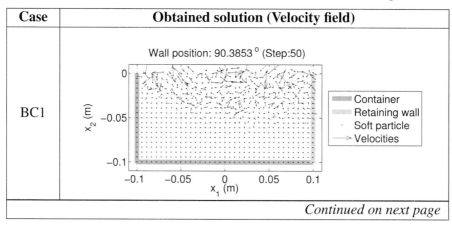
	Continued on next page

Table 6.3 : (*continued*)

Case	Obtained solution (Velocity field)
BC2	
BC3	
BC4	
BC5	

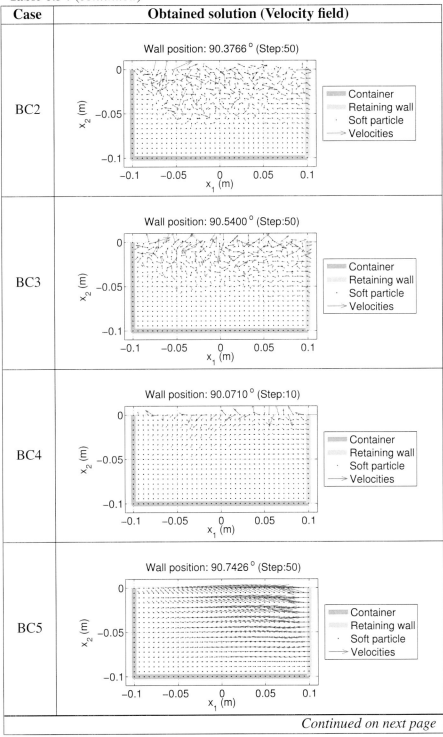

Continued on next page

Table 6.3 : (*continued*)

Case	Obtained solution (Velocity field)
BC6	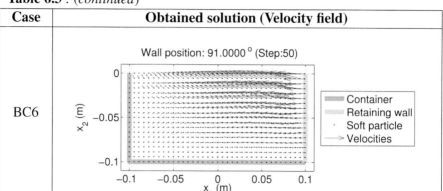

In cases BC5 and BC6, the velocity fields during outward tilts (tilting away from sand) are less smooth, compared with the those during inward tilts (tilting against sand), as shown in Fig. 6.7. For cases BC5 and BC6, the wall tilts 1° in approximately 50 steps. Thus, 600 steps are approximately three cycles of tilt. After two to three cycles of tilt, the velocity fields in cases BC5 and BC6 become chaotic, as shown in Fig. 6.8. The simulation of case BC5 was aborted manually due to divergence even with a very small Δt. The simulation of BC6 was aborted due to the existence of 'NaN' or 'Inf' in the Jacobian matrix. During these tilts, no shear band formed in the simulation. Also, in the experiments with a tilt amplitude of 1°, no shear bands nor slip surfaces were observed within the first 5 cycles of tilt.

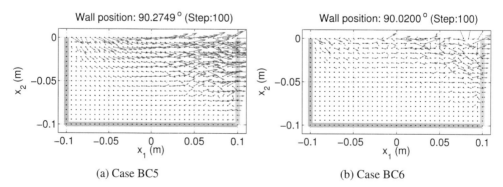

(a) Case BC5   (b) Case BC6

Figure 6.7: Velocity fields of cases BC5 and BC6 after the first outward wall tilt, at step 100. In these two cases, the wall tilts 1° in about 50 steps.

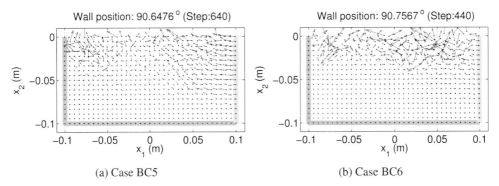

(a) Case BC5   (b) Case BC6

Figure 6.8: Velocity fields of cases BC5 and BC6, at the step before the simulations were aborted.

6.3.1 Larger number of neighbors

In this sub-section, the number of neighbors in the simulations with boundary conditions shown in Table 6.1 is increased. The examples of neighbors are shown in Fig. 6.9.

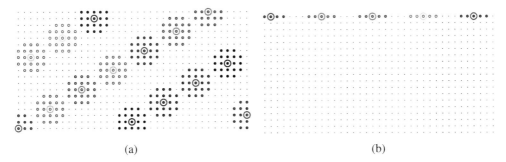

(a)   (b)

Figure 6.9: Examples of larger numbers of neighbors used for the evaluation of (a) $\mathbf{L}$ and $\nabla \cdot \mathbf{T}$, and (b) unit vector $\mathbf{n}_f$ normal to the free surface. Particles with the same color are neighbors of the circled ones.

The obtained velocity fields of the first time increment are shown in Table 6.4. In cases BC1 through BC4, the velocity fields are chaotic. The velocity fields in cases BC5 and BC6, although not quite smooth, are relatively more rational. With increasing simulation steps (see Fig. 6.5), the velocity fields of BC1 through BC4 remain chaotic. Simulations of cases BC1, BC2, BC4, and BC6 were aborted due to divergence even with a very small Δt. BC3 was aborted due to the appearance of 'NaN' or 'Inf' in the Jacobian matrix. BC6 was manually aborted since the velocity field became chaotic.

Table 6.4: Obtained velocity field at the first time increment for the simulations with larger supports. The cases correspond to the boundary conditions shown in Table 6.1 and simulation set-ups detailed in Section 6.2.

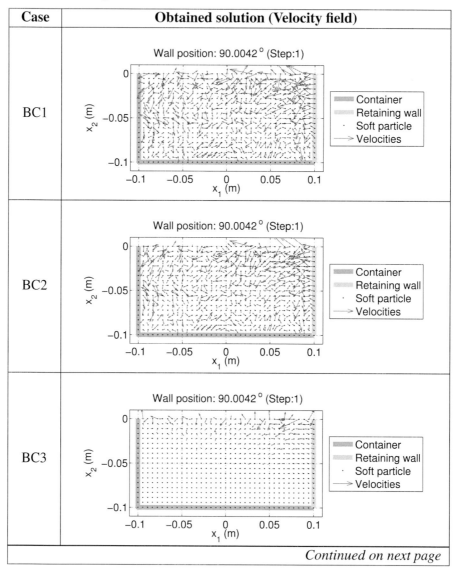

Case	Obtained solution (Velocity field)
BC1	
BC2	
BC3	

Continued on next page

Table 6.4 : (*continued*)

Case	Obtained solution (Velocity field)
BC4	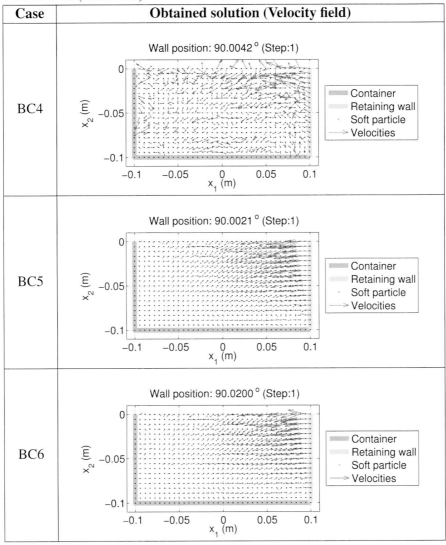
BC5	
BC6	

Table 6.5: Obtained velocity field of simulations with a larger number of the neighbors, at the simulation step shortly before the simulations were aborted. The cases correspond to the boundary conditions shown in Table 6.1.

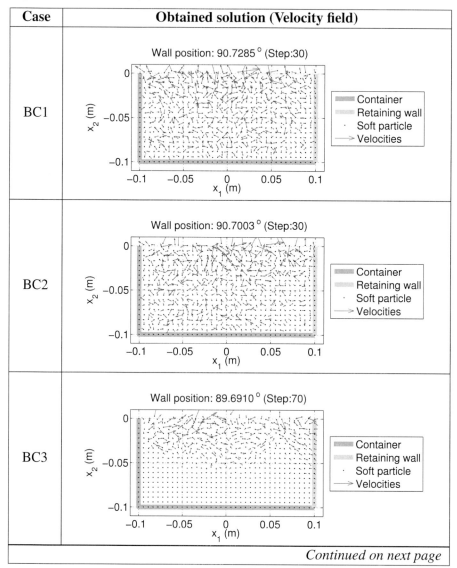

Case	Obtained solution (Velocity field)
BC1	
BC2	
BC3	

Continued on next page

Table 6.5 : (*continued*)

Case	Obtained solution (Velocity field)
BC4	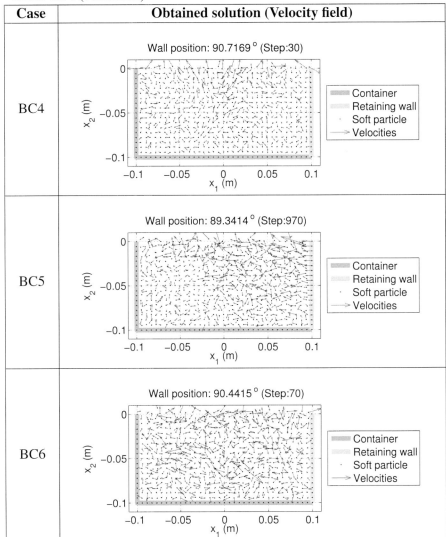
BC5	
BC6	

6.3.2   Second order polynomials

In this sub-section, second order polynomials, eq. 3.2, are used for the simulations with boundary conditions shown in Table 6.1. More neighbors are now used for the evaluation of spatial derivatives, as shown in Fig. 6.9. The simulation results in terms of velocity fields are shown in Table 6.6.

The velocity fields of cases BC1, BC2, and BC4 are chaotic already at the first time increment. The velocity fields of these cases remain chaotic in later simulation steps. The simulations of these cases lasted no more than 50 steps and were aborted due to divergence even with a very small Δt.

The velocity field of case BC3 is very similar to the initial solution (see Fig. 6.5), as Δt was decreased to 10^{-8} in order to obtain converge. Five steps later, Δt was decreased to 10^{-15} in order to obtain converge, and hence the simulation was aborted manually. The obtained velocity fields of the six steps are identical to the initial guess of the velocity field due to the usage of a extremely small value for Δt from the first simulation step.

The simulation of case BC5 lasted for 1400 steps, but the velocity field became chaotic after 60 steps, as shown in Fig. 6.10a. Case BC6 was aborted after 71 steps due to the usage of a very small Δt. Its velocity field became chaotic after 20 steps, as shown in Fig. 6.10b. Compared with the velocity field in the simulations with first order polynomials that became chaotic in much later steps (approximately after 250 steps for case BC5 and 100 steps for case BC6).

Table 6.6: Obtained velocity fields at the first time increment for the simulations using second order polynomials. The cases correspond to the boundary conditions shown in Table 6.1 and simulation set-ups detailed in Section 6.2.

Case	Obtained solution (Velocity field)
BC1	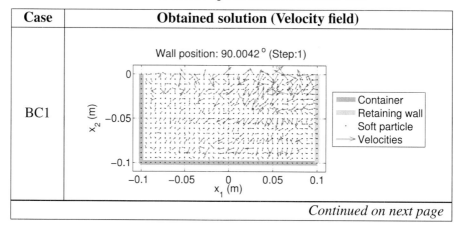
	Continued on next page

Table 6.6 : (*continued*)

Case	Obtained solution (Velocity field)
BC2	
BC3	
BC4	
BC5	

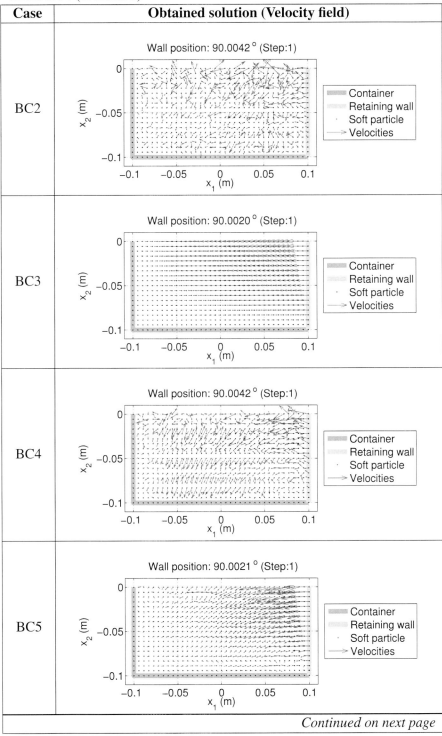

Continued on next page

Table 6.6 : (*continued*)

Case	Obtained solution (Velocity field)
BC6	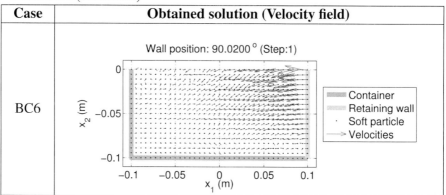

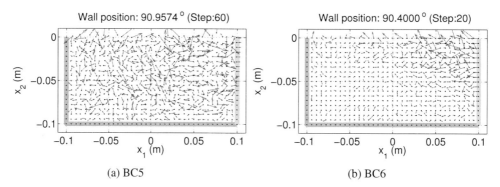

(a) BC5   (b) BC6

Figure 6.10: Chaotic velocity fields obtained for case BC5 and BC6 at steps 60 and 20, respectively.

6.3.3 More rigorous stop criterion

A more rigorous stop criterion $\epsilon \leq 10^{-6}$ in the numerical solver was also applied. It was found out that some simulations are aborted earlier, as this small tolerance $\epsilon \leq 10^{-6}$ could not be achieved, and decreasing the stop criterion from 10^{-2} to 10^{-6} does not improve the results (in terms of the smoothness of velocity fields obtained by the numerical solver used herein).

6.4 Summary

Simulations with six set-ups of boundary conditions shown in Table 6.1 have been carried out. Despite all the efforts to implement various boundary conditions and to use different equilibrium equations for the free surface, the deformation of sand induced by a cyclic tilting wall could not be successfully modeled with SPARC. The obtained velocity fields are non-smooth or become chaotic within one cycle of wall tilts.

Chapter 7

Summary, Limitations and Outlook

7.1 Summary

This work focuses on the development of the soft particle code (SPARC) for numerical simulations. The current version of SPARC can deal with quasi-static problems. In SPARC, the unknowns are velocities of particles. The spatial derivatives are evaluated with the polynomial interpolation/approximation method using information stored at nearby particles. The way the spatial derivatives are evaluated is simple and is one of the main research points of SPARC. It makes SPARC different from other numerical methods. Thereafter, a system of nonlinear equations is numerically solved. Among the solvers implemented in this work, the arc-length method performs the best. The simulation examples shown adopt the nonlinear and inelastic constitutive model, barodesy. The code has been tested in the simulations of the oedometer test, triaxial test and biaxial test. Despite of the simplicity of SPARC, it has been shown in Chapter 5 that the obtained results are reasonable and comparable with, e.g., element test simulations, a finite-element simulation, and the X-ray analysis of a deformed triaxial sample. In addition, shear band formation can be modeled. The parameter study (Chapter 3) reveals for biaxial tests that SPARC performs the best with a regular array of particles and first order polynomials with a minimum number of neighbors. Assuming homogeneous initial fields in biaxial tests, homogeneous deformation or stress fields can be modeled using a regular array of particles. When an irregular array of particles is used, inhomogeneous deformation (strain localization) appears at a smaller axial strain. In biaxial test, after the formation of shear bands (Appendix F), the velocity field becomes less homogeneous and the deformation field becomes more complicated. However, it is difficult to argue the correctness of this shear band formation process. Furthermore, model tests using the zigzag apparatus were conducted (Appendix J). The motion of sand grains in the backfill exhibits closed trajectories. Simulations of the laboratory model tests with an initially horizontal free surface were carried out (Chapter 6). Despite all the efforts to implement various boundary conditions (for the free surface and the frictional boundary), the deformation of sand induced by a cyclic tilting wall could not be successfully modeled.

7.2   Limitations

This framework is proposed in attempting to overcome a problem of mesh-based methods such as FE methods. However, the equilibrium equations are still (e.g. in FE methods) difficult to be solved after strain localization takes place. Convergence problems raise from the highly nonlinear and inelastic nature of the constitutive model, barodesy, and from the inhomogeneous deformation. Although considerable efforts have been made to solve the convergence problems, convergence is still not possible after shear bands have appeared a while. Based on the results presented in this work, the following limitations of the proposed method have been noticed:

- When the neighbors are updated from one time step to another, the obtained stress-strain curve in the simulation of the triaxial test exhibits a small jump (see Fig. 4.9 in Section 4.2.5.2). This indicates that changes of the number of neighbors have an significant influence on the system. This might result from the way the spatial derivatives are evaluated in SPARC. The obtained velocity gradient at a particle can be different for different neighbors in support.

- SPARC is capable of modeling shear band formation in biaxial test simulations using a regularly array of particles and first-order polynomial. With an irregular array of particles, assuming initial homogeneous fields, the convergence is more difficult (more iterations are required) and strain localization appears much earlier. The initialization of the shear band formation is thus dependent on the particle arrangement. It also has been shown that the deformation fields obtained using an irregular array of particles are less homogeneous. The contours of the deformation field or the possible shear band patterns are complicated, which can lead to convergence problems. Besides, second order polynomials result in a larger width of the shear band. After the formation of one or more shear bands, convergence is not possible. These all evident that the first order polynomial interpolation/approximation method is strongly influenced by the configuration of the particles and the order of polynomial for the evaluation of the spatial derivatives.

- Simulations in Chapter 6 reveal that SPARC can not model the 'zig-zag'-tests (Appendix J).

7.3 Outlook

Based on the limitations of the current version of SPARC, some recommendations are given in the following.

1. Adaptive time integration scheme using substeps has been suggested in [18, 17] for the automatic selection of a substep size, achieved by controlling maximum error of stress in each substep. Such an error control also improves convergence in the FE method. The adaptive time integration scheme is hence suggested to be implemented.

2. Several scattered data approximation methods (e.g. weighted least-squares method, radial basis function, etc.) are known (e.g. [15]). The polynomial interpolation/approximation adopted in SPARC has made the framework simple, but seems to have a limitation on the simulation of large deformations. Since the evaluation of the spatial derivatives is relevant, it is worth trying out various approximation methods on the proposed framework. In addition, in weighted approximation methods (e.g. weighted least-squares methods), distant neighbors have a fewer influence to the derivatives. That is, the impact to the system when a particle enters or leaves a support is considerably reduced, although the weighted approximation methods are generally computationally more expensive.

3. SPARC is coded in Matlab. The computation efficiency has been significantly improved by compiling the code with Matlab Coder and the code is parallelized making use of the Matlab Parallel Computing Toolbox. In order to simulate problems with a large number of particles, it is suggested to translate the code in another programming language and to develop the algorithm for parallelization computing utilizing a large number of processors. However, parallelization alone can be a demanding task as pointed out by Hoover [25], the efforts required to reprogram algorithms for parallelization "can easily be an order of magnitude larger than that required for developing a single processor algorithm".

4. The numerical solver is relevant for solving the system of equations. This work has shown that the Newton's method and the Levenberg-Marquardt method encounter convergence problem around the peak of the stress-strain curve or during shear band formation. Even with the arc-length method, no solution could be determined when the accumulated strains become large. The Broyden – Fletcher – Goldfarb – Shanno (BFGS) method has been proved to work well in FE methods [1]. In the BFGS method, the stiffness matrix is constructed based on the one used in the previous time increment. Although the convergence difficulties might also rise from searching strong solutions or other facts, it might be worth trying out the BFGS method.

5. The simulation results of the 'zig-zag'-tests in Chapter 6 have shown that the velocity fields obtained with frictional boundary implemented (i.e. cases BC1 through BC4 in Table 6.1) are chaotic, whereas the velocity fields obtained in the simulations without the frictional boundary (i.e. cases BC5 and BC6 in Table 6.1), although not smooth, are not chaotic and the simulations were obtained for more cycles of tilt. This implies that the frictional boundary condition might make the system of equations more complicated. The reason why the chaotic and non-smooth velocity fields were obtained is still unclear. It is suggested to investigate the frictional boundary condition.

6. For dynamic problems, explicit schemes should be developed in SPARC. The stable fourth-order Runge-Kutta time integration scheme, as has been applied to dynamical problems in [25, 6], could be a good choice.

7. The numbers of particles in an unit spatial volume are similar everywhere in the study domain. As the space is discretized, more particles are required when higher resolution is required. It is still unclear, how to adjust the size of supports when the number of particles in an unit spatial volume significantly varies.

Bibliography

[1] *Bathe K.J.* (1996), *Finite Element Procedures*, Prentice Hall, Upper Saddle River, New Jersy.

[2] *Belystchko T.*, *Liu Y.Y.* and *Gu L.* (1994), Element-Free Galerkin Methods, *International Journal for Numerical Methods in Engineering*, **37**: pp. 229–256.

[3] *Belytschko T.*, *Krongauz Y.*, *Organ D.*, *Fleming M.* and *Krysl P.* (1996), Meshless methods:An overview and recent developments, *Comput. Methods Appl. Mech. Engrg.*, **139**: pp. 3–47.

[4] *Beuth L.* (2012), *Formulation and Application of a Quasi-Static Material Point Method*, PhD thesis, Universität Stuttgart.

[5] *Blanc T.* (2011), *The Runge-Kutta Tylor-SPH Model, A New Improved Model for Soil Dynamics Problems*, PhD thesis, Universitdad Politécnica de Madrid.

[6] *Blanc T.* and *Pastor M.* (2011), A stabilized Smoothed Particle Hydrodynamics, Taylor–Galerkin algorithm for soil dynamics problems, *INTERNATIONAL JOURNAL FOR NUMERICAL AND ANALYTICAL METHODS IN GEOMECHANICS*, dOI: 10.1002/nag.1082.

[7] *Bobryakov A.*, *Kosykh V.* and *Revuzhenko A.* (1990), Temporary structures during the deformation of granular media, Fiziko-Tekhnicheskie Problemy Razrabotki Poleznykh Iskopaemykh, **1**: pp. 29–39, translated by Plenum Publishing Corporation.

[8] *Brackbill J.* and *Ruppel H.* (1986), FLIP: A method for adaptively zoned, particle-in-cell calculations in two dimensions, *Journal of Computational Physics*, **65**.

[9] *Butcher J.* (1996), A history of Runge-Kutta methods, *Applied Numerical Mathematics*, **20**: pp. 247–260.

[10] *Cheang L.* and *Matlock H.* (1983), Static and cyclic lateral load tests on instrumented piles in sand, Tech. rep., The Earth Technology Corporation, Long Beach, California.

[11] *Chen C.H.* and *Kolymbas D.* (2015), Sand eddies induced by cyclic tilt of a retaining wall, *Acta Geotechnica*, pp. 1–12.

[12] *Chen C.H.*, *Schneider-Muntau B.* and *Kolymbas D.*, Soft Particle Code, Advantages and Limitations in Comparison to Finite Element Method., in progress.

[13] *Coetzee C.J.* (2003), *The Modelling of Granular Flow Using the Particle-in-Cell Method*, PhD thesis, University of Stellenbosch.

[14] *Cuellar P.*, *Maessler M.* and *Ruecker W.* (2009), Ratchcting convcctive cells of sand grains around offshore piles under cyclic lateral loads, *Granular Matter*, **11**: pp. 379–390.

[15] *Fasshauer G.E.* (2007), *Meshfree Approximation Methods with Matlab*, World Scientific Publishing Co. Pte. Ltd.

[16] *Fellin W.* (2000), *Rütteldruckverdichtung als plastodynamicshes Problem*, p. 159, A.A.Balkema/Rotterdam/Brookfield.

[17] *Fellin W.*, *Mittendorfer M.* and *Ostermann A.* (2009), Adaptive integration of constitutive rate equations, *Computers and Geotechnics*, **36**: p. 698–708, doi:10.1016/j.compgeo.2008.11.006.

[18] *Fellin W.* and *Ostermann A.* (2002), Consistent tangent operators for constitutive rate equations, *INTERNATIONAL JOURNAL FOInt. J. Numer. Anal. Meth. Geomech.*, **26**: p. 1213–1233, dOI: 10.1002/nag.242.

[19] *Fries T.P.* and *Matthies H.G.* (2004), *Classification and Overview of Meshfree Methods*.

[20] *Gilbert J.R.*, *Moler C.* and *R. S.* (1992), Sparse matrices in matlab: Design and Implementation, *SIAM Journal on Matrix Analysis and Applications.*, **13-1**: pp. 333 – 356.

[21] *Gingold R.A.* and *Monaghan J.* (1977), Smoothed Particle Hydrodynamics: Theory and Application to Nonspherical Stars, *Monthly Notices of the Royal Astronmical Society*, **181**: pp. 375–389.

[22] *Harlow F.* (1964), *The Particle-in-Cell Computing Method for Fluid Dynamics in Fundamental Methods in Hydrodynamics.*, pp. 319–345, Experimental Arithmetic, High-Speed Computations and Mathematics, Academic Press.

[23] *Harlow F.H.* and *E. W.J.* (1965), Numerical Calculation of Time-Depende Viscous Incompressible Flow of Fluid with Free surface, *The Physics of Fluids*, **8**(12): pp. 2182–2189.

[24] *Higgins B.G.*, *Arc length Continuation Methods: An Introduction*, Department of Chemical Engineering & Materials Science University of California, Davis. http://www.ekayasolutions.com/ech256/ECH256ClassNotes/ArcLengthContinuation.pdf, teaching handout.

[25] *Hoover W.G.* (2006), *Smooth Particle Applied Mechnics*, World Scientific, iSBN 981-270-002-1.

[26] *Keller H.B.* (1986), *Lectures on Numerical Methods in Bifurcation Problems.*, Springer-Verlag, Berlin, Heidelberg, New York, Tokyo.

[27] *Kolymbas D.* (1972), Master's thesis, Universität Karlsruhe, Deutschland.

[28] *Kolymbas D.* (1999), *Introduction to Hypoplasticity*, A. A. Balkema.

[29] *Kolymbas D.* (2011), Thematische Spezialisierung 4 - Theoretische Bodenmechanik, course handout. Universität Innsbruck, in German.

[30] *Kolymbas D.* (2012), Barodesy: a new constitutive frame for soils., *Géotechnique Letters*, **2**: pp. 17–23.

[31] *Kolymbas D.* (2012), Barodesy: a new hypoplastic approach, *Int. J. Numer. Anal. Meth. Geomech.*, **36**: pp. 1220–1240, dOI: 10.1002/nag.1051.

[32] *Kolymbas D.* (2012), Barodesy as a novel hypoplastic constitutive theory based on the asymptotic behaviour of sand., *geotechnik*, **35 (3)**: pp. 189–197.

[33] *Kreyszig E.* (1999), *Advanced Engineering Mathematics*, WILEY.

[34] *Kum O.*, *Hoover W.G.* and *Posh H.A.* (1995), Viscous conducting flows with smooth-particle applied mechanics, *Phys. Rev. E*, **109**: pp. 67–75.

[35] *Levenberg K.* (1944), A Method for The Solution of Certain Non-linear Problems in Least-squares, *Quarterly of Applied Mathematics*, **2**: pp. 164–168.

[36] *Li S.* and *Liu W.K.* (2002), Meshfree and particle methods and their applications, *Appl. Mech. Rev. American Society of Mechanical Engineers.*, **55**(1): pp. 1–34, review articles.

[37] *Li S.* and *Liu W.K.* (2004), *Meshfree Particle Methods*, Springer Berlin Heidelberg New York, iSBN 978-3-540-22256-9.

[38] *Liu G.* and *Liu M.* (2007), *Smoothed Particle Hydrodynamics - a meshfree particle method*, World Scientific, iSBN 981-238-456-1.

[39] *Lu Y.Y.*, *Belytschko T.* and *Gu L.* (1994), A new implementation of the element free Galerkin method, *Comput. Methods Appl. Mech. Engrg.*, **113**: pp. 397–414.

[40] *Lucy L.B.* (1977), A Numerical Approach to the Testing of the Fission Hypothesis, *The Astronomical Journal*, **82**: pp. 1013–1024.

[41] *Mabssout M.*, *Herreros M.* and *Pastor M.* (2006), Wave propagation and localization problems in saturated viscoplastic geomaterials, *International Journal for Numerical Methods in Engineering*, **68**: pp. 425–447.

[42] *Mabssout M.* and *Pastor M.* (2003), A Taylor-Galerkin algorithm for shock wave propagation and strain localization failure of viscoplastic continua, *Computer Methods in Applied Mechanics and Engineering*, **192**: p. 955–971.

[43] *Mabssout M.* and *Pastor M.* (2003), A two-step Taylor-Galerkin algorithm applied to shock wave propagation in soils, *International Journal for Numerical and Analytical Methods in Geomechanics*, **27**: pp. 685–704.

[44] *Mabssout M.*, *Pastor M.*, *Herreros M.* and *Quecedo M.* (2006), A Runge-Kutta Taylor-Galerkin scheme for hyperbolic systems with source terms. Application to shock wave propagation in viscoplastic geomaterials, *International Journal for Numerical and Analytical Methods in Geomechanics*, **30**: p. 1337– 1355.

[45] *Marquardt D.W.* (1963), An Algorithm for Least-squares Estimation of Nonlinear Parameters, *Journal of the Society for Insustrial and Applied Mathematics*, **11**(2): pp. 431–441.

[46] *Medicus G.*, *Fellin W.* and *Kolymbas D.* (2012), Barodesy for clay, *Géotechnique Letters 2*, pp. 17–23.

[47] *Nayroles B.*, *Touzot G.* and *Villon P.* (1992), Generalizing the finite element method: diffuse approximation and diffuse elements, *Computational Mechanics*, **10**: pp. 307–318.

[48] *Papadopoulos A.* and *Manolopoulos Y.* (1997), Performance of Nearest Neighbor Queries in R-Trees, in *Proceedings of the 6th International Conference on Database Theory.*, LNCS 1186, pp. 394–408, Delphi, Greece.

[49] *Posh H.A.*, *Hoover W.G.* and *Kum O.* (1995), Steady-state shear flows via nonequilibrium molecular dynamics and smooth-particle applied mechanics, *Phys. Rev. E*, **52**: pp. 1711–1719.

[50] *Silpa-Anan C.* and *Hartley R.* (2008), Optimized kd-trees for fast image descriptor matching, in *Proceedings of the IEEE Conference on Computer Vision and Pattern Recognition*.

[51] *Stock A.* (2013), *A High-Order Particle-in-Cell Method for Low Density Plasma Flow and the Simulation of Gyrotron Resonator Devices*, PhD thesis, Universität Stuttgart.

[52] *Sulsky D.*, *Zhou S.J.* and *Schreyer H.L.* (1995), Application of a particle-in-cell method to solid mechanics, *Comput. Phys. Commun.*, **87**: pp. 236–252.

[53] *Thielicke W.* and *Stamhuis E.J.*, PIVlab - Time-Resolved Digital Particle Image Velocimetry Tool for MATLAB, *Program downloadable at http://pivlab.blogspot.co.at/*.

[54] *Vaidya P.* (1989), An O(n log n) Algorithm for the All-Nearest-Neighbors Problem, *Discrete and Computational Geometry*, **4**(1): pp. 101–115.

[55] *Wagner P.* (2006), *Skriptum der Vorlesung - Mathematik A*, Chap. V. Differentialrechnung in mehreren Variablen, Innsbruck Universität, in German. http://mat1.uibk.ac.at/wagner/skripten.html.

[56] *Wubs F.* (2009), *Numerical Bifurcation Analysis of Large Scale System.*, Chap. 4, University of Groningen, lecture notes in Applied Mathematics. Acdamic year 2009-2010.

[57] *Zheng B.*, *Xu J.*, *Lee W.C.* and *Lee D.L.* (2006), Grid-partition index: a hybrid method for nearest neighbor queries in wireless location-based services, *The VLDB Journal*, **15**(1): pp. 21–39.

Appendix A

Generation of a Point Cloud

The method for the generation of a point cloud used in this work is demonstrated in a two dimensional space.[1] The idea is to look for a *hole* with radius r_h among points. A hole is a circular space in which no points are present. Once a hole is identified, a point is added to its center.

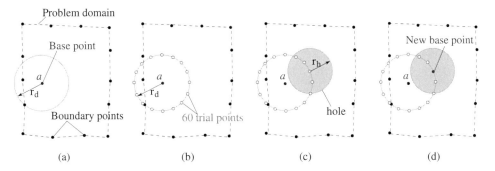

Figure A.1: Illustration of the procedure for the generation of a point cloud. The solid points are soft particles.

Procedure:

1. Generate *boundary points* which define a closed domain, i.e. black solid points in Fig. A.1a. Note that only solid points illustrated in Fig. A.1 are used as soft particles. Then, generate the first point (point a in Fig. A.1a) arbitrarily located in the domain. a is considered as the *base point*.

2. Use the *base point* as the center and generate 60 *trial points* on a circle around it (with radius r_d), as shown in Fig. A.1b. Trial points are not used as soft particles.

3. The search of a hole begins at one of the *trial points*. A hole exists when there are no mass points (soft particles) located within the radius r_h around a *trial point*, as shown in Fig. A.1c. In such case, a mass point is added at that location and it is assigned as the new *base point*, as shown in Fig. A.1d.

---

[1]The procedure documented in this Appendix is based on a suggestion by Jörg Kuhnert (Fraunhofergesellschaft, Kaiserslautern).

4. Repeat the previous two steps until no more holes can be found at *trial points* ever generated.[2]

Note that r_h must be slightly smaller than r_d. $r_h = 0.7r_d$. The generated point cloud is shown in Fig. A.2.

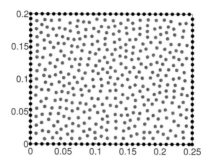

Figure A.2: Illustration of the generated point cloud.

---

[2]For this reason, the coordinates of all trial points ever generated must be stored during the generation of a point cloud.

Appendix B

Sub-Domain k-Nearest Neighbor Search

Neighbor search is one of the essential parts of a meshless method, as the spatial derivatives at particles are evaluated[1] using their neighbors. Two neighbor search methods are investigated in this study: The fixed-radius search method and the k-nearest neighbor search. The latter is documented herein.

The k-nearest neighbor search (k-nn search) method is to find out the nearest k neighbors of a point $\mathbf{x}_{(i)}$[2]. The naive idea of the k-nn search for a query point is to:

1. calculate the (Euclidean) distances from that query to all other points in the problem;
2. sort all the distances; and
3. take the nearest k points as neighbors of that query point.

However, computationally it becomes more expensive when a large number of points are used. Space partitioning has been applied in neighbor search methods (e.g. index-tree structure[3] methods and grid-partition index method[4]) to reduce search time. It serves as a filter that maintains an index structure[5] and the candidate neighbors are selected based on the index structure.

The neighbor search problem in SPARC is an all-nearest-neighbors problem [54], meaning that all points in a problem are query points. As a fast and accurate k-nn search method is required in SPARC, the *sub-domain k-nn search* method adopting space partitioning along with some filter techniques is proposed.

---

[1] A *particle* is a point which carries physical information such as density, void ratio, stress tensor etc. Instead, a *point* is referred to as a set of coordinates in an n-dimensional space.

[2] The point is indicated with brackets "()"

[3] such as kd-trees (a multidimensional indexing method) [50], R-tree [48] (a method that uses 'R'ectangles to group nearby data) , etc.

[4] Zheng et al. [57] introduced the grid-partition index which is constructed based on the Voronoi diagram. The space is divided into cells so that the search space is reduced. In a grid cell, information about candidates is provided through the Voronoi diagram.

[5] "Traditional nearest-neighbor search is based on two basic indexing approaches: objective-based indexing and solution-based indexing. The former is constructed based on the locations of data objects using some distance heuristics on object locations. The latter is built on a pre-computed solution space." [57]

B.1   Sub-domain k-nn search

B.1.1   Domain partitioning by sorting coordinates of points

The first step of the sub-domain k-nn search method is to partition the problem domain. This is carried out by sorting the coordinates of the points. The partitions are termed sub-domains. Thereafter, the points are assigned into the sub-domains. The procedure is demonstrated by the example shown in Fig. B.1: There are 1000 points in the 2D domain. The domain is partitioned into 4×6 sub-domains. The points are firstly sorted according to the x-components of their coordinates. Secondly, every 250 points[6] are numerically assigned the same index indicating that they are in the same sub-domain, as shown in Fig. B.1b. Finally, in every sub-domain, the points are sorted based on the y-components of their coordinates and, thereafter, every 50 points[7] are assigned into the individual sub-domains (Fig. B.1c).

This can be extended to 3D problems, as shown in Fig. B.2. These two examples show that sorting the particles using their coordinates, followed by indexing them, naturally results in partitioning.

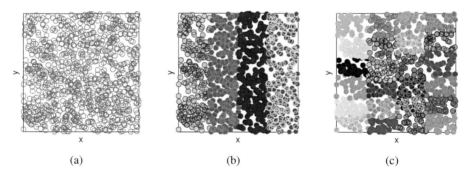

| (a) | (b) | (c) |

Figure B.1: Space partitioning by sorting points' coordinates. (a) Given 1000 particles in space. (b) Points in space are sorted using x-components of their coordinates and every 250 points are assigned into the individual sub-domains. (c) Points in each sub-domain are sorted using y-components of their coordinates and every 50 points are assigned into the individual sub-domains.

---

[6]This quantity is roughly estimated by dividing the number of points in space by the desired number of sub-domains in x-direction.

[7]This quantity is roughly estimated by dividing the number of points in the sub-domain under evaluation by the desired number of partitions in y-direction.

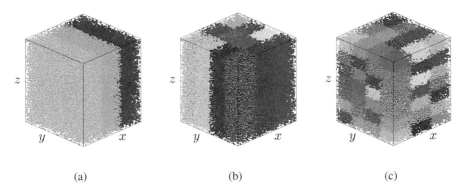

(a)   (b)   (c)

Figure B.2: A 3D example of space partitioning by sorting coordinates. There are 100,000 points in space. The problem domain is partitioned into $3 \times 4 \times 7$ sub-domains

B.1.2 Indexing sub-domains

Every sub-domain is indexed by a set of representative coordinates, determined by the mean values of the coordinates of all particles in each sub-domain.

B.1.3 Neighbor search of sub-domains

Since the k-nn search in the proposed method is carried out using only particles collected in the same and the nearby sub-domains, the neighbors of every sub-domain must be determined prior to k-nn search.

A sub-domain is said to be a neighbor of the query point's sub-domain if the distance between the representative coordinates of the two sub-domains is smaller than a prescribed length r_s. Note that very few computations are needed in this step, as the number of the sub-domains is much smaller compared with the number of points in the study domain.

B.1.4 k-nn search

Searching the k-nearest neighbors of a query is carried out in the following procedure, as shown in Fig. B.3. Consider a query point located in sub-domain A:

1. The points which are in sub-domain A and in A's neighbors are collected (Fig. B.3a).

2. For the query point:

(a) The points which are far away by either of the following criteria are filtered out :

- $a > r_t$ and $b > r_t$ (Fig. B.3b)
- $a + b > r_t$ (Fig. B.3c)

where r_t is a suitable prescribed length.

(b) For the remaining points (i.e. the solid dots in Fig. B.3b and B.3c), which satisfy either $\{a < r_t \text{ and } b < r_t\}$ or $\{a+b < r_t\}$, their Euclidean distances ($r_{(j)} = \sqrt{a^2 + b^2}$) to the query point are calculated.

(c) All the calculated distances $r_{(j)}$ are sorted and the nearest k points are taken as the nearest neighbors.

It is noted that r_t should not be too small such that the quantity of remaining points are fewer than k. In SPARC, $r_s \approx 2r_t$ is applied and the criterion, $\{a > r_t \text{ and } b > r_t\}$, is employed to filter out points (i.e. Fig. B.3b) prior to the computations of the Euclidean norm.

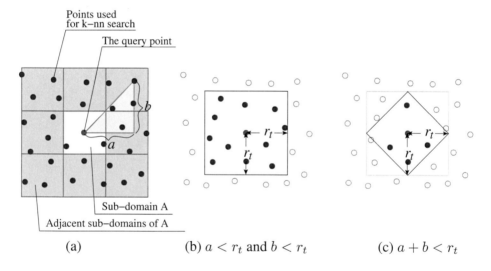

| (a) | (b) $a < r_t$ and $b < r_t$ | (c) $a + b < r_t$ |

Figure B.3: Illustration of (a) k-nn search using particles in the nearby sub-domains, (b) filter rule: $a < r_t$ and $b < r_t$, and (c) filter rule: $a + b < r_t$.

B.2   Efficiency of the sub-domain k-nn search method

The method is written in Matlab and compiled using Matlab Coder to a binary MEX file. The Matlab built-in function, knnsearch, is used for reference:

```
[n,d]=knnsearch(x,y,'k',12,'distance','minkowski','p',5);
```

This function performs a k-nn search between points, whose coordinates are stored in the matrix **x**; and the coordinates of query points stored in matrix **y**.

Study cases:
Fig. B.4 shows the configuration of points used in study case A, containing 3359 points. Study case B consists of 29,887 points.[8] The parameters used in the proposed method and the corresponding computation times for both cases are listed in Tables B.1 and B.2.

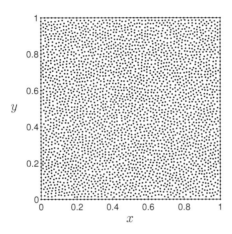

Figure B.4: Configuration of points in study case A.

k	Matlab k-nn (sec)	Sub-domain k-nn (sec)	Number of partitions in x	Number of partitions in y	r_t	r_s
10	0.0669	0.0687	10	10	0.080	0.160
10	0.0665	0.0511	12	12	0.070	0.140
10	0.0660	0.0413	14	14	0.070	0.140
20	0.0674	0.0668	10	10	0.080	0.160
30	0.0666	0.0513	12	12	0.070	0.140

Table B.1: computation time of case A (Fig. B.4).

The results indicate that the computation times of Matlab built-in function and those of the proposed method are not much affected by the k value. However, the efficiency of the proposed method is based on the set-up of the parameters (i.e. r_t and r_s). While looking for the k-nearest neighbors of a query, only the points in the adjacent sub-domains are used. As a result, the size of the sub-domains and the number

---

[8]The configuration of study case B is not shown in here because the large number of points black out the figure.

k	Matlab k-nn (sec)	Sub-domain k-nn (sec)	Number of partitions in x	Number of partitions in y	r_t	r_s
10	0.6271	0.5113	30	30	0.070	0.140
10	0.6301	0.3779	35	35	0.060	0.120
10	0.6082	0.3127	40	40	0.055	0.110
50	0.6277	0.5528	30	30	0.070	0.140
50	0.6282	0.7206	25	25	0.080	0.160

Table B.2: computation time for case B.

of nearby sub-domains being used affect the amount of computations. With fine-tuned parameters used in the proposed method, the amount of computations can be minimized, resulting in less computation time.

It is particularly noted that, compared to the computational efforts for the computation of the stiffness matrix in the Newton's method, the computational costs for neighbor search are negligible.

Recommendations for parameters in sub-domain k-nn method
The following recommendations are made after testing the proposed method.

1. The quantity of points in a sub-domain shall be larger than the k value (to make sure the nearest neighbors are really the nearest ones).

2. The number of the nearby sub-domains, in which all points are used for k-nn search, is affected by r_s. r_s is, thus, selected in a way such that a sub-domain is surrounded by other sub-domains. This can be achieved by plotting the sub-domain and its adjacent sub-domains, as illustrated in Fig. B.5b, B.5c and B.5d. In Fig. B.5b, too few nearby sub-domains are taken. If any of the adjacent sub-domains are not taken, the found k-nearest points are not necessarily the nearest ones. On the contrary, too many nearby sub-domains are taken in Fig. B.5d. This results in unnecessary extra computations. The r_s used in Fig. B.5c is considered to be a good choice in the study case, as all adjacent sub-domain are included.

3. $r_t \approx r_s/2$ has been found to be a good choice for r_t.

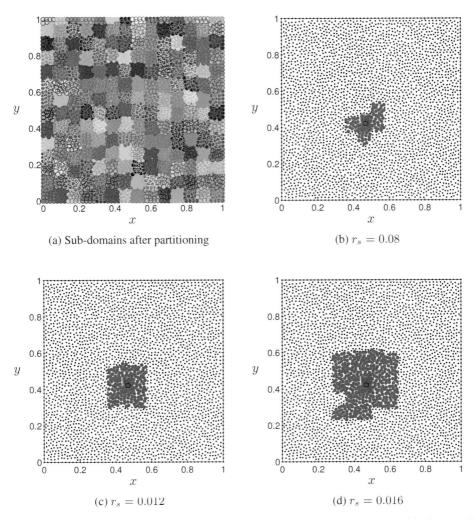

(a) Sub-domains after partitioning

(b) $r_s = 0.08$

(c) $r_s = 0.012$

(d) $r_s = 0.016$

Figure B.5: Illustration of the selection of r_s for case A (Fig. B.4a) with the number of partitions equal to 14 in both x- and y-directions. Consequently, there are 18 points in each sub-domain. (a) Sub-domains. Points in the same sub-domain are marked with the same color. In (b), (c) and (d), the red solid dot is the query point. Red circles are particles in the query point's sub-domain. Blue dots are particles in the nearby sub-domains of the query point. They are used for the k-nn search.

Appendix C

Mapping Vector m

In this appendix, it is to show how to use the mapping method described in Section 2.6.3 (page 17) to obtain (1) the residuals of the system of equations ($\mathbf{y}$) from the matrix of residuals at all particles (Ξ) and (2) the list of unknowns ($\mathbf{u}$) from the matrix of velocity vectors at all particles $\mathbb{V}$.

It must be noted that all the known variables u or $\mathbf{u}$ are referred as $u^{t+\Delta t}$ or $\mathbf{u}^{t+\Delta t}$.

Consider a problem with totally 5 ($=n_\mathrm{p}$) particles in 3D space (with axes x_1, x_2 and x_3). Based on the dimension of the problem, we define the matrix numeration as

$$\mathcal{N} = \begin{bmatrix} 1 & 6 & 11 \\ 2 & 7 & 12 \\ 3 & 8 & 13 \\ 4 & 9 & 14 \\ 5 & 10 & 15 \end{bmatrix} \tag{C.1}$$

This 5×3 matrix stores the position in a vector denoted by integers $1, 2, \cdots, 3n_\mathrm{p}$.

Given the matrix $\mathcal{D}^\mathrm{of}$ that defines the degrees of freedom of a problem:

$$\mathcal{D}^\mathrm{of} = \begin{matrix} x_1\ x_2\ x_3 & \leftarrow \text{components} \\ \begin{bmatrix} 1 & 0 & 0 \\ 0 & 1 & 1 \\ 1 & 1 & 0 \\ 1 & 1 & 0 \\ 0 & 0 & 0 \end{bmatrix} & \begin{matrix} \leftarrow \text{degrees of freedom of particle 1} \\ \leftarrow \text{degrees of freedom of particle 2} \\ \leftarrow \text{degrees of freedom of particle 3} \\ \leftarrow \text{degrees of freedom of particle 4} \\ \leftarrow \text{degrees of freedom of particle 5} \end{matrix} \end{matrix} \tag{C.2}$$

where integer 1 denotes the corresponding component is an unknown variable in $\mathbb{V}$ (eq. 2.44, page 17), and 0 denotes a parameter with known values in $\mathbb{V}$.

Given computed residuals (see Section 2.43, page 16):

$$\Xi = \begin{array}{ccc} x_1 & x_2 & x_3 \quad \leftarrow \text{components} \end{array}$$

$$\Xi = \begin{bmatrix} 20 & 0 & 0 \\ 0 & -100 & -120 \\ 30 & 80 & 15 \\ 40 & 0 & 0 \\ 6 & -60 & -20 \end{bmatrix} \begin{array}{l} \leftarrow \text{residuals of particle 1} \\ \leftarrow \text{residuals of particle 2} \\ \leftarrow \text{residuals of particle 3} \\ \leftarrow \text{residuals of particle 4} \\ \leftarrow \text{residuals of particle 5} \end{array} \qquad \text{(C.3)}$$

Thus we obtain the mapping vector

$$\mathbf{m} = \begin{bmatrix} 1 & 3 & 4 & 7 & 8 & 9 & 12 \end{bmatrix}^{\mathrm{T}} \qquad \text{(C.4)}$$

and the degrees of freedom of the system $d^{\mathrm{of}} = 7$.

Consequently, the residuals in $\mathbf{y}$ ($\mathbf{y} = y_{i_e}$, $i_e = 1 \cdots d^{\mathrm{of}}$) of the system of equations $\mathbf{y}(\mathbf{u}^{t+\Delta t}) = \mathbf{0}$ are obtained via the mapping vector $\mathbf{m}$ as follows (see eq. 2.48, page 18). It consists of the residuals in Ξ where the corresponding components in $\mathbb{V}$ are unknown:

$$\mathbf{y} = y_{i_e} = \begin{bmatrix} y_1 \\ y_2 \\ y_3 \\ y_4 \\ y_5 \\ y_6 \\ y_7 \end{bmatrix} = \begin{bmatrix} \Xi(\mathbf{m}(1)) \\ \Xi(\mathbf{m}(2)) \\ \Xi(\mathbf{m}(3)) \\ \Xi(\mathbf{m}(4)) \\ \Xi(\mathbf{m}(5)) \\ \Xi(\mathbf{m}(6)) \\ \Xi(\mathbf{m}(7)) \end{bmatrix} = \begin{bmatrix} \Xi(& 1 &) \\ \Xi(& 3 &) \\ \Xi(& 4 &) \\ \Xi(& 7 &) \\ \Xi(& 8 &) \\ \Xi(& 9 &) \\ \Xi(& \underbrace{12}_{\text{position in } \Xi} &) \end{bmatrix} = \begin{bmatrix} 20 \\ 30 \\ 40 \\ -100 \\ 80 \\ 0 \\ -120 \end{bmatrix} \qquad \text{(C.5)}$$

The unknowns in $\mathbf{u}$ ($\mathbf{u} = u_{i_u}$, $i_u = 1 \cdots d^{\mathrm{of}}$) of the system of equations $\mathbf{y}(\mathbf{u}) = \mathbf{0}$ are also obtained via the mapping vector $\mathbf{m}$ as follows. Given velocity of all particles: (For unknown variables, initial guessed values are given.)

$$
\begin{array}{ccc}
x_1 & x_2 & x_3 & \leftarrow \text{components}
\end{array}
$$

$$
\mathbb{V} =
\begin{bmatrix}
\boxed{0} & 0 & 0 \\
1 & \boxed{1} & \boxed{-0.2} \\
\boxed{2.5} & \boxed{2} & 0.3 \\
\boxed{0.5} & \boxed{-1} & 2 \\
1.2 & 2 & -2
\end{bmatrix}
\begin{array}{l}
\leftarrow \mathbf{v}^{t+\Delta t} \text{ of particle 1} \\
\leftarrow \mathbf{v}^{t+\Delta t} \text{ of particle 2} \\
\leftarrow \mathbf{v}^{t+\Delta t} \text{ of particle 3} \\
\leftarrow \mathbf{v}^{t+\Delta t} \text{ of particle 4} \\
\leftarrow \mathbf{v}^{t+\Delta t} \text{ of particle 5}
\end{array}
\qquad \text{(C.6)}
$$

Analog to the calculation procedure of $\mathbf{y}$, the list of unknowns in $\mathbf{u} = u_{i_u}, i_u = 1 \cdots d^{\text{of}}$ is computed (i.e. eq. 2.50, page 18):

$$
\mathbf{u} = u_{i_u} =
\begin{bmatrix}
u_1 \\ u_2 \\ u_3 \\ u_4 \\ u_5 \\ u_6 \\ u_7
\end{bmatrix}
=
\begin{bmatrix}
\mathbb{V}(\mathbf{m}(1)) \\
\mathbb{V}(\mathbf{m}(2)) \\
\mathbb{V}(\mathbf{m}(3)) \\
\mathbb{V}(\mathbf{m}(4)) \\
\mathbb{V}(\mathbf{m}(5)) \\
\mathbb{V}(\mathbf{m}(6)) \\
\mathbb{V}(\mathbf{m}(7))
\end{bmatrix}
=
\begin{bmatrix}
\mathbb{V}(& 1 &) \\
\mathbb{V}(& 3 &) \\
\mathbb{V}(& 4 &) \\
\mathbb{V}(& 7 &) \\
\mathbb{V}(& 8 &) \\
\mathbb{V}(& 9 &) \\
\mathbb{V}(& \underbrace{12}_{\text{position in } \mathbb{V}} &)
\end{bmatrix}
=
\begin{bmatrix}
0 \\ 2.5 \\ 0.5 \\ 1 \\ 2 \\ -1 \\ -0.2
\end{bmatrix}
\qquad \text{(C.7)}
$$

While computing the Jacobian matrix, an increment (δu) is firstly added to one of the components of $\mathbf{u}$ (eq. 2.58, page 21) and then the residuals in $\mathcal{E}$ (eq. 2.42, page 16) are computed. Thus, we must map the changed component back into $\mathbb{V}$ so that we know which velocity component of which particle has been changed. Say u_5 is changed to $\bar{u}_5 = u_5 + \Delta u$ while evaluating $\frac{\partial (\cdot)}{\partial u_5}$ in the Jacobian matrix:

$$
\leadsto i_u = 5 \leadsto \mathbf{m}(i_u) = \mathbf{m}(5) = 8 \qquad \text{(C.8)}
$$
$$
\leadsto \mathbb{V}(\mathbf{m}(i_u)) = \mathbb{V}(8) = u_5 + \delta u \qquad \text{(C.9)}
$$

and 8th component in $\mathbb{V}$ becomes:

$$
\mathbb{V} =
\begin{bmatrix}
0 & 0 & 0 \\
1 & 1 & -0.2 \\
2.5 & 2 + \Delta u & 0.3 \\
0.5 & -1 & 2 \\
1.2 & 2 & -2
\end{bmatrix}
\qquad \text{(C.10)}
$$

Appendix D

Determination of Normal Vectors on a Surface

In SPARC, the surface particles are particles representing the boundaries. The normal vectors on the surface particles are determined by the following procedure. In this appendix, only surface particles are considered (Fig. D.1a).

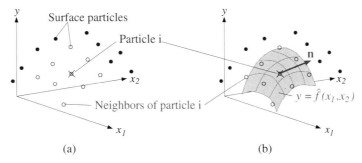

Figure D.1: (a) Surface particles. (b) Constructed surface $y = f(x_1, x_2)$ at particle i and the unit normal vector computed by means of the surface function y.

1. Surface construction at particle i:
   Now, we consider one particle, particle i (Fig. D.1a). The surface of particle i is constructed using its neighbors, as described in eqs. (2.18) through (2.23). As a result, the approximated surface function y at particle i can be written as:

   $$y = \hat{f}(x_1, x_2) = a_1 x_1 + a_2 x_2 \tag{D.1}$$

   where the coefficients a_1 and a_2 are obtained using eq. (2.23).

2. Evaluation the unit normal vector $\mathbf{n}$ at particle i:
   The unit normal vector $\mathbf{n}$ at particle i on the surface is evaluated by computing the gradient of constructed surface y of particle i:

   $$\mathbf{N} = \left(-\frac{\partial y}{\partial x_1}, -\frac{\partial y}{\partial x_2}, 1 \right) = (-a_1, -a_2, 1) \tag{D.2}$$

   $$\mathbf{n} = \mathbf{N}/|\mathbf{N}| \tag{D.3}$$

129

3. Repeat the previous two steps to obtain the normal vectors of all surface particles

The following three functions (normal_vector_example, get_normal_vector and get_neighbor_list) are provided to assist the comprehension of the above procedure. It follows the aforementioned procedure.

```
function normal_vector_example()
% ================================================================
% % Demonstration of computing normal vector using
% %     surface particles (MAIN PROGRAM):
% ================================================================
    clc, clear all, close all
    % GENERATION OF SURFACE PARTICLES:   (Stored in x1, x2 and y)
    spacing = 0.2;   v = -2:spacing:1;
    [x_1,x_2] = meshgrid(v); % get grid data
    y_ = x_2*x_1./10 .* 3.^(-2*x_1.^2 - .5*x_2.^2);
    x1 = x_1(:); x2 = x_2(:); y=y_(:);

    % COMPUTING NORMAL VECTORS at PARTICLE i
    %    USING ITS NEIGHBORS
    n = zeros(length(y),3);
    for i = 1: length(y)
        % Neighbor search
        support_radius = spacing*1.7;
        xyz = [x1, x2]; no_of_p = length(x1);
        [neighbor_list,no_of_neighbors]=get_neighbor_list(xyz, ...
            support_radius,no_of_p, i);
        % Computing the normal vector at particle i
        n(i,1:3)=get_normal_vector(xyz(neighbor_list,:), ...
            no_of_neighbors,y(neighbor_list), xyz(i,:),y(i));
    end

    % FIGURE GENERATION
    figure(292)
    mesh(x_1,x_2,y_), hold on % Plot the surface particles
    quiver3(x1,x2,y,n(:,1),n(:,2),n(:,3)) % plot the normal vectors
    daspect([1,1,1]), axis tight, view(-30,34), zoom(1.2)
    xlabel('x_1'), ylabel('x_2'), zlabel('y')
end
```

The following function computes the normal vector at a particle by constructing the surface using eqs. (2.18) through (2.23). The surface is re-constructed using information stored at the neighbors of the particle.

```
function [n]=get_normal_vector(xyz_neighbors,no_of_neighbors,Y ...
    ,xyz_particle,Y_particle)
% This function computes the normal vector n at
%     position 'xyz_particle':
    % === INPUT =====================================================
    % xyz_neighbors = coordinates of neighbors (? x 2)
    % no_of_neighbors = the total number of neighbors of the particle
    %    'xyz_particle' (1 x 1)
    % Y = values of neighbors at xyz_neighbors
    % xyz_particle = coordinates of the particle at which the normal
    %    vector to the surface is computed. (2 x 1)
```

```
% Y_particle = the value of the particle at which the normal
%   vector is computed (1 x 1)
% === OUTPUT ====================================================
% n = the normal vector to the surface (1x3)

% CONSTRUCT SURFACE f= a_1*x_1 * a_2*x_2
xyz_shifted = zeros(no_of_neighbors,2);
% Shift coordinate
xyz_shifted(1:no_of_neighbors,1)=xyz_neighbors(:,1)-xyz_particle(1);
xyz_shifted(1:no_of_neighbors,2)=xyz_neighbors(:,2)-xyz_particle(2);
% Evaluate the coefficients a_1 and a_2 in f
a = xyz_shifted\(Y-Y_particle);

% EVALUATE NORMAL VECTOR n
n=zeros(1,3); n(1)=-a(1); n(2)=-a(2); n(3)=1;
n(1:3)= n(1:3)./norm(n);
end
```

The following function carries out the fixed-radius neighbor search method for the considered particle.

```
function [neighbor_list,no_of_neighbors]=get_neighbor_list(xyz, ...
                    support_radius,no_of_p, particle_index)
% This function finds the neighbors of a particle with index
% 'particle_index':
    % === INPUT ====================================================
    % xyz= coordinates of particles (? x 2)
    % support_radius = the radious of support (1 x 1)
    % no_of_p = the total number of particles (1 x 1)
    % particle_index = the index of the particle whose
    %   neighbors is searched herein
    % === OUTPUT ===================================================
    % neighbor_list = list of neighbor (? x 1)
    % no_of_neighbors = the number of neighbors of the particle
    %     'particle_index'

    points_series =  1:no_of_p;

    % The position of the particle
    center= [xyz(particle_index,1)   xyz(particle_index,2) ];
    % Particles outside this box defined by [xub, xlb], [yub, ylb],
    % and [zub, zlb] will be removed (ub=UpperBound, lb=LowerBound ):
    xub = center(1) + support_radius;  xlb - center(1)  support_radius;
    yub = center(2) + support_radius;  ylb = center(2) - support_radius;
    % Remove particles which are not neighbors
    logic_neigh = xyz(:,1) <= xub & xyz(:,1) >= xlb;
    logic_neigh = xyz(:,2) <= yub & xyz(:,2) >= ylb & logic_neigh;
    % Get the indices of the candidate particles which could be neighbors
    xyz_neigh_index = points_series(logic_neigh);
    % Get the coordinate of the candidate particles
    x_ = xyz(xyz_neigh_index,1);   y_ = xyz(xyz_neigh_index,2);
    % Check which candidate particles are located in the radius
    r_ = sqrt((x_-center(1)).^2 + (y_-center(2)).^2);
    t_ = r_<=support_radius;
    % Update the neighbor indices of neigbors found
    xyz_neigh_index = xyz_neigh_index(t_);
    % Save the index of neighboring points to a list
    no_of_neighbors= length(xyz_neigh_index);
```

```
    % Store the list of neighbors
    neighbor_list = zeros(no_of_neighbors,1);
    neighbor_list(1:no_of_neighbors)=xyz_neigh_index;
end
```

The computed unit normal vectors using the code above are shown Fig. D.2a. The surface particles are linked (i.e. the grid) to visualize the surface. The arrows are unit vectors normal to the surface at the particles.

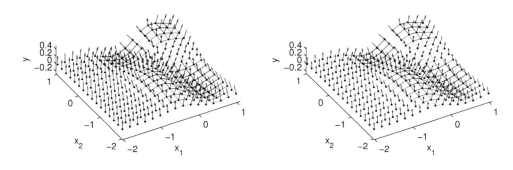

(a) Interpolation/approximation - SPARC   (b) With Matlab built-in function 'gradient'

Figure D.2: Computed normal vectors (denoted by arrows) on the surface particles (black dots). The particles are linked (the mesh) to visualize the surface.

For comparison, a Matlab code using the built-in function `gradient` to evaluate the normal vectors is given below. The obtained vectors are shown in Fig. D.2b.

```
% NORMAL VECTOR EVALUATED USING Matlab
clc, clear all, close all
%generate data points from y=f(x_1,x_2)
spacing = 0.2;   v = -2:spacing:1;
[x_1,x_2] = meshgrid(v); % get grid data
y = x_2*x_1./10 .* 3.^(-2*x_1.^2 - .5*x_2.^2); % y = f(x_1,x_2)

% Evaluation of the normal vectors
[px_1,px_2] = gradient(y,.2,.2); % get the gradient of the function y
figure(294)
mesh(x_1,x_2,y), hold on % plot the surface
%plot the normal vectors
quiver3(x_1,x_2,y,-px_1,-px_2,zeros(size(px_1))+1)
plot3(x_1,x_2,y,'k.','MarkerFaceColor','k')
daspect([1,1,1]), axis tight, view(-30,34), zoom(1.2)
xlabel('x_1'), ylabel('x_2'), zlabel('y')
```

Appendix E

Sparse Matrix

A sparse matrix is a matrix in which the ratio of the number of non-zero elements to the number of total elements in a matrix is very low. In SPARC, the stiffness matrix is sparse, as indicated by their sparsity patterns shown in Fig. E.1.

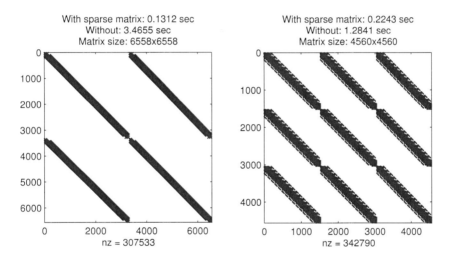

(a) 2D simulation of a biaxial test.　　　　(b) 3D simulation of a triaxial test.

Figure E.1: Visualization of the sparsity pattern of the stiffness matrix in the Newton's method. 'nz' denotes the number of non-zeros elements in the stiffness matrices. (a) The sparsity pattern of the stiffness matrix in a 2D simulation of the biaxial test. Computation times of the matrix inversion using sparse matrix is 0.1312 sec and without using sparse matrix is 3.4655 sec. (b) The sparsity pattern of the stiffness matrix in a 3D simulation of the triaxial test. Computation times of the matrix inversion using sparse matrix is 0.2243 sec and without using sparse matrix is 1.2841 sec.

The computation time for inverting the stiffness matrix (in Newton's method, eq. 2.61) is shorter when the stiffness matrix is stored as sparse matrix. Computing using sparse matrices is effective for the following reasons:

1. Zeros are not stored, hence the storage is reduced. Such a way of storing data is called sparse storage.

2. The amount of computations are significantly reduced as only non-zero elements are processed.

A useful documentation regarding the implementation of sparse matrices in Matlab can be found in [20].

Appendix F

Evolution of Shear Bands

The evolution of the shear bands obtained in Section 3.2.2 is shown in the following figure.

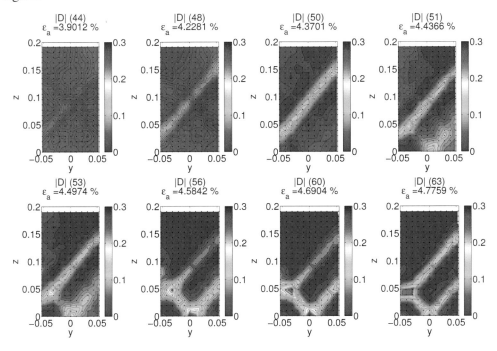

Figure F.1: Evolution of shear band at selected simulation steps. The contours of $|\mathbf{D}|$ are illustrated. A weak zone with a higher void ratio is implemented (see Fig. 3.12). The number in the brackets indicate the simulation step. ε_a is the axial strain of the sample. (*continued*)

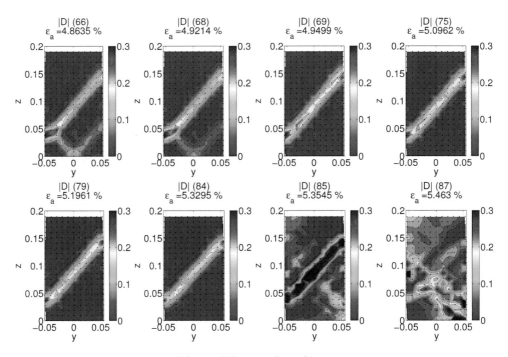

Figure F.2: (continued.)

Appendix G

An Example of Scaling a Matrix

In the LEVENBERG-MARQUARDT method, MARQUARDT [45] suggested to scale $\mathbf{A}$ matrix using its main diagonal of components (i.e. eqs. 2.70 and 2.72). Let $\mathbf{A}^*$ be the scaled matrix of $\mathbf{A}$:

$$\mathbf{A}^* = A_{ij}^* \qquad \text{(eq.2.70)}$$

$$A_{ij}^* = \frac{A_{ij}}{\sqrt{A_{ii}}\sqrt{A_{jj}}} \qquad \text{(eq.2.72)}$$

This appendix is to provide a calculation example for this operation. Let

$$\mathbf{A} = \begin{bmatrix} 2 & 5 & 9 \\ 1 & 7 & 3 \\ 4 & 2 & 1.2 \end{bmatrix}$$

$\mathbf{A}^*$ is calculated as follows:

$$\mathbf{A}^* = A_{ij}^* = \begin{bmatrix} \frac{2}{\sqrt{2}\sqrt{2}} & \frac{5}{\sqrt{2}\sqrt{7}} & \frac{9}{\sqrt{2}\sqrt{1.2}} \\ \frac{1}{\sqrt{7}\sqrt{2}} & \frac{7}{\sqrt{7}\sqrt{7}} & \frac{3}{\sqrt{7}\sqrt{1.2}} \\ \frac{4}{\sqrt{1.2}\sqrt{2}} & \frac{2}{\sqrt{1.2}\sqrt{7}} & \frac{1.2}{\sqrt{1.2}\sqrt{1.2}} \end{bmatrix}$$

Appendix H

An example for the explanation of arc-length parameterization

The following example[1] explains the definition of "the arc-length parameterization" or "the parameterization according to arc length". The position vector γ parameterized with time $t \in \mathbb{R}$ is a path in space of $\gamma \in \mathbb{R}^n$, as shown in Fig. H.1a.

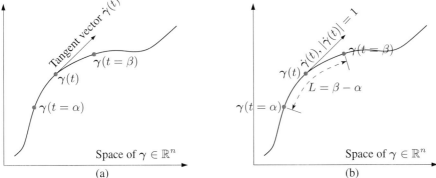

(a)   (b)

Figure H.1: Illustration of arc-length. (a) A tangent vector on a curve $\gamma(t)$ in the space $\mathbb{R}^n$. (b) Taking $|\dot{\gamma}(t)| = 1$ during time $t \in [\alpha, \beta]$ resuts in $L = \beta - \alpha = \Delta t$.

The tangent vector of the curve at time t is the velocity at $\gamma(t)$ on the path, i.e. the time derivative of $\gamma(t)$:

$$\dot{\gamma}(t) = \frac{d\gamma}{dt} = [\dot{\gamma}_1(t), \dot{\gamma}_2(t), \ldots \dot{\gamma}_n(t)]^T \tag{H.1}$$

The speed is the length of the velocity : $|\dot{\gamma}(t)|$. If one walks along the path from the time $t = \alpha$ to $t = \beta$, the total walking distance L (length of the path, or the *arc length*) is the time integration of the speed [55]:

$$L = \int_\alpha^\beta |\dot{\gamma}| \, dt = \int_\alpha^\beta \sqrt{\dot{\gamma}_1(t)^2 + \dot{\gamma}_2(t)^2 + \cdots + \dot{\gamma}_n(t)^2} \, dt \tag{H.2}$$

Walking faster or taking longer for the walk leads to a longer walking distance L.

---

[1] Communicated by Professor Peter Wagner, Innsbruck University

Particularly, if one walks at a constant speed of 1 (i.e. $|\dot{\gamma}(t)| = 1$) for a period of time $t \in [\alpha, \beta]$, the walking distance L (Fig. H.1b) is

$$L = \int_{\alpha}^{\beta} 1 \, dt = \beta - \alpha = \Delta t \tag{H.3}$$

In this particular case, γ is said to be parameterized according to arc length.

Appendix I

Derivation of the Equation of a Plane

This appendix provides the derivation of the plane used in the pseudo-arc-length method (Section 2.7.4.4). Given $\boldsymbol{\gamma}^t \in \mathbb{R}^{N+1}$ and the solution $\dot{\boldsymbol{\gamma}}^{t-\Delta t,\,0}$ determined in time step $t - \Delta t$, as shown in Fig. I.1. $\boldsymbol{\gamma}_A$ is away from $\boldsymbol{\gamma}^t$ along $\dot{\boldsymbol{\gamma}}^{t-\Delta t,\,0}$:

$$\boldsymbol{\gamma}_A = \boldsymbol{\gamma}^t + \dot{\boldsymbol{\gamma}}^{t-\Delta t,\,0} \Delta t \tag{I.1}$$

If $\boldsymbol{\gamma}$ is a point on the plane with normal vector $\mathbf{n} = \dot{\boldsymbol{\gamma}}^{t-\Delta t,\,0}/|\dot{\boldsymbol{\gamma}}^{t-\Delta t,\,0}|$ passing $\boldsymbol{\gamma}_A$, the equation of the plane reads:

$$(\boldsymbol{\gamma}_A - \boldsymbol{\gamma}) \cdot \mathbf{n} = 0 \tag{I.2}$$
$$\rightsquigarrow (\boldsymbol{\gamma}^t + \dot{\boldsymbol{\gamma}}^{t-\Delta t,\,0} \Delta t - \boldsymbol{\gamma}) \cdot \mathbf{n} = 0$$
$$\rightsquigarrow \dot{\boldsymbol{\gamma}}^{t-\Delta t,\,0} \Delta t \cdot \mathbf{n} - (\boldsymbol{\gamma} - \boldsymbol{\gamma}^t) \cdot \mathbf{n} = 0$$
$$\rightsquigarrow \dot{\boldsymbol{\gamma}}^{t-\Delta t,\,0} \cdot \mathbf{n} - \frac{(\boldsymbol{\gamma} - \boldsymbol{\gamma}^t)}{\Delta t} \cdot \mathbf{n} = 0$$
$$\rightsquigarrow \dot{\boldsymbol{\gamma}}^{t-\Delta t,\,0} \cdot \mathbf{n} - \dot{\boldsymbol{\gamma}}^t \cdot \mathbf{n} = 0 \tag{I.3}$$

Eq. (I.3) is used in the pseudo-arc-length method replacing the arc-length condition $|\dot{\boldsymbol{\gamma}}^t| = 1$ (eq. 2.85, Section 2.7.4, page 29).

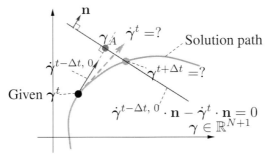

Figure I.1: Construction of the plane $\dot{\boldsymbol{\gamma}}^{t-\Delta t,\,0} \cdot \mathbf{n} - \dot{\boldsymbol{\gamma}}^t \cdot \mathbf{n} = 0$ which replaces the arc-length condition.

Appendix J

Sand Eddies Induced by Cyclic Tilt of a Retaining Wall

The content of the this appendix has been prepared as a journal paper [11] accetped by Acta Geotechnica.

The cyclic tilt of a retaining wall induces a peculiar motion in the backfill (sand), which exhibits closed trajectories (eddies). In this paper the motion of the backfill has been optically traced and analyzed by means of PIV (particle image velocimetry), also known as DIC (digital image correlation). The results are of importance for cyclically loaded structures (e.g, piles for off-shore structures) and can also serve to test numerical simulations of large deformation.

J.1 Introduction

As first observed by BOBRYAKOV et al. [7], the cyclic tilt of a retaining wall induces large deformations in the backfill consisting of dry sand. CUÉLLAR et al. [14] applied cyclic loading to a pile embedded in saturated sand and found out that soil grains mixed beneath the free surface. Both of them directly [7] and indirectly [14] observed the convective motion of grain flow in the backfill. To study this motion in detail, a model test apparatus was built at the Division of Geotechnical and Tunnel Engineering of the University of Innsbruck (see Section J.2.1). The grain motion was traced by digital photographs and analyzed by means of PIV (see Section J.2.4).

We found that some grains track loops of quasi-elliptical form, as shown in Fig. J.2. With a series of tests we analyzed the influence of the following parameters (see Section J.2.2):

- amplitude of tilt, and

- geometry of the backfill.

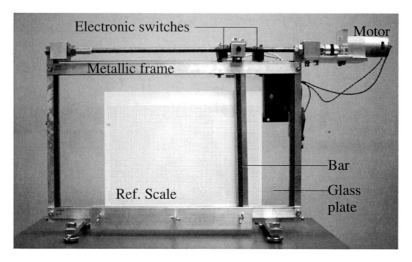

Figure J.1: Experimental apparatus.

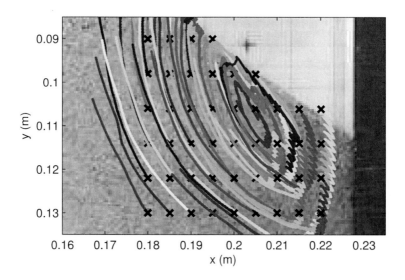

Figure J.2: Quasi-elliptical form of grain paths. Each line illustrates a trajectory starting at a location marked with '×'.

J.2 Experimental setup

J.2.1 Apparatus

The apparatus is shown in Fig. J.1. An aluminum bar serves as the retaining wall and is hinged at the bottom to enable tilt. The retaining wall was tilted by a motor and two electronic switches altering the directions of the drive of the motor. The bar and backfill are contained within two scratch-resistant glass plates fixed on a stiff metal frame with a spacing of 12 mm. The tilt amplitude α of the retaining wall can be set by adjusting the positions of the electronic switches. For reference, a one-millimeter graph paper is affixed on the back-side glass. The backfill material consists of dry sand with a grain size distribution shown in Fig. J.3.

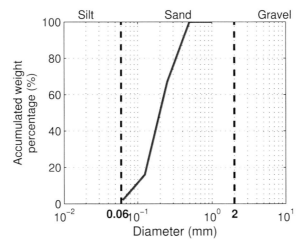

Figure J.3: Grain size distribution of test sand.

J.2.2 Test series

The conducted tests are summarized in Table J.1. The geometry parameters listed in Table J.1 are illustrated in Fig. J.4. The tilt speed of the wall for $\alpha = 1°$ is 6 seconds per cycle and that for $\alpha = 3°$ is 18 seconds per cycle.

J.2.3 Preparation of the backfill

The preparation of the backfill is as follows: the wall is set to its vertical position with $\alpha = 0°$. Dry sand is pluviated using a funnel and keeping a constant falling height. The funnel is moved slowly back and forth to make a homogeneous backfill.

Table J.1: Test series.

Name	height $h2/h1$ (cm)	surface inclination $\beta(°)$	tilt amplitude $\alpha(°)$	number of cycles
Test 1	14.0/14.0	0	3	60
Test 2	16.1/7.8	21	3	171
Test 3	-/8.5	34	1	236
Test 4	14.3/14.3	0	1	281

*The initial density of sand can be hardly determined. Evaluating the total volume and the mass of sand, we found for Test 1 a value for density of $\rho = 1.5$ g/cm^3 and we assume that this value is valid for all tests.

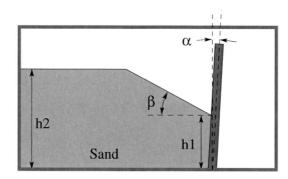

Figure J.4: Definition of the parameters in Table J.1.

The free surface was inclined by various angles, see Fig. J.4 and Table J.1. The initial tilt direction, either inwards or outwards, was found to have no influence on the results.

J.2.4   Particle image velocimetry

In usual sand, the individual grains have various colors and form with their arrangement characteristic patterns. Comparing the patterns at times t and $t + \Delta t$, we can identify clusters of grains and their motion. Of course, the deformation alters the form of each cluster. However, for small Δt the alteration is small and the individual clusters can be identified by digital correlation analysis. This enables tracing the motion of the clusters and, hence, the motion of the grains. Consecutively tracing of this motion makes it possible to find the trajectories (paths) of grains over large time lapses. In our tests, a digital camera *Sony* $\alpha 35$ with a resolution of 3568×2368 pixels is used. The camera was placed in a distance of $0.50 - 0.60$ m from the apparatus and the illumination was achieved with a halogen spotlight (500 Watt, J-Type, IP54) placed on the upper right or left corner behind the camera avoiding the direct reflection of the spotlight itself. The camera was leveled by water-levels and oriented perpendicular to the glass plate. The open-source software PIVLab [53] was used to conduct the PIV analysis. It provides a friendly graphical user interface for image input and for PIV analysis, and has a variety of output formats and built-in functions for post-analysis. Besides, the user may easily mask out specific regions with arbitrary shapes in the images to avoid unnecessary calculations in the PIV analysis. The computation is also parallelized which allows an analysis with larger number of high-resolution images to be carried out in significantly shorter time.

J.3   Results

J.3.1   Evolution of free surface

The evolution of the free surfaces in our experiments is illustrated in Fig. J.5. In Tests 1, 2 and 4, the subsidence of the free surfaces was observed. It was found to occur only near the wall. This area is termed "conical soil depression zone" by CHEANG AND MATLOCK [10], and has been observed in their quasi-static cyclic tests. CUÉLLAR et al. [14] also observed such a phenomenon in their model test, along with two heaves at the edge of a conical soil depression zone on both sides of the pile in the loading direction. In addition, BOBRYAKOV et al. [7] and CUÉLLAR et al. [14] used the position of the free surface as a criterion to distinguish the motions of the backfill. To explain the hereafter used notion of *global heave* and *local heave*, we refer to Fig. J.6. A *global heave* appears at the edge of the depression zone whereas a *local heave* indicates a temporary tiny one within the depression zone.

Tests 1 and 4 (Figs. J.5a and J.5b) started with a horizontal free surface, the same configuration as in CUÉLLAR et al. [14]. The tilt amplitudes were $\alpha = 3°$ and $\alpha = 1°$, respectively. Test 2 started with a surface inclination angle of $21°$, similar to the experiment carried out by BOBRYAKOV et al. [7]. In the early cycles of these experiments before the free surface subsided to a steady position, the backfill slipped into the empty space created when the retaining wall tilted outward. Therefore, the surface near the wall dropped a bit, forming a slope. This motion corresponds to the failure mechanism of active earth pressure, as shown in Fig. J.7a. When the retaining wall was pushed against the backfill, about 2/3 of the total length of the surface close to the wall rose slightly and a global heave began to form. This type of movement is associated with passive earth pressure, as shown in Fig. J.7b. The pressure response caused by the backfill acting on the wall can be found in BOBRYAKOV et al. [7].

With increasing number of cycles, the surface inclination became steeper and the surface near the wall continued to sink until the surface reached a stationary posi-

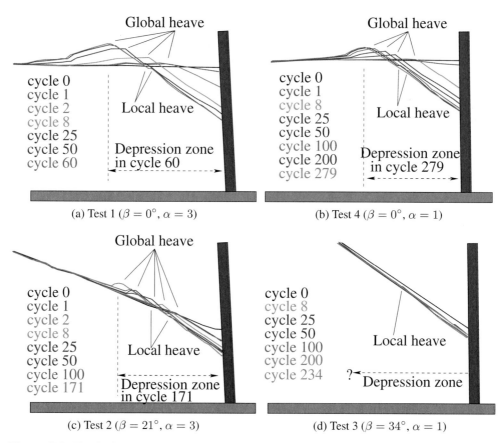

(a) Test 1 ($\beta = 0°$, $\alpha = 3$)

(b) Test 4 ($\beta = 0°$, $\alpha = 1$)

(c) Test 2 ($\beta = 21°$, $\alpha = 3$)

(d) Test 3 ($\beta = 34°$, $\alpha = 1$)

Figure J.5: Evolution of the free surface. Free surfaces at various cycles are marked with different colors.

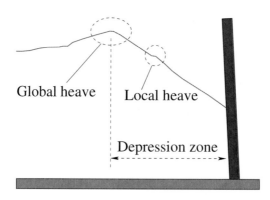

Figure J.6: Definition of *global heave*, *local heave*, and *depression zone* from a free surface.

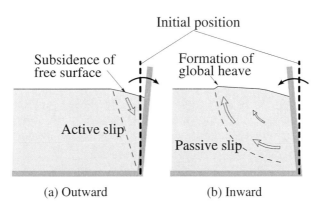

Figure J.7: Illustration of the formation of slope and heave(s) induced by (a) outward and (b) inward wall tilts.

tion. At this position, the slope angle is $35°$, measured by the average surface inclination over the depression zone. This angle is termed the *critical inclination angle* throughout this paper. Thereafter, from cycle to cycle, the global heave migrated progressively away from the wall to the left and became bigger (e.g. see cycles 8, 25, 50 in Fig. J.5a and cycles 25, 50, 100 in Fig. J.5b); the surface sank deeper. The results also show that the size of the global heave is proportional to the tilt amplitude.

After a number of cycles, the alteration of the surface in shape and position significantly retarded (e.g. cycle 25 in Fig. J.5a and cycle 100 in Fig. J.5b). In this situation, the surface reached a stationary position although minor changes continued over cycles, e.g. the migration local heaves, the appearance of new local heaves and the migration of the global heave. BOBRYAKOV et al. [7] used this phenomenon as a criterion to divide the motion of the backfill in their experiment into firstly a non-stationary phase, followed by a stationary phase characterized as "steady-state material density and unchanging specimen surface". The same phenomenon has also been observed by CUÉLLAR et al. [14], with the depression of the surface reaching a steady depth "within the first thousands of cycles". In [14], two distinct phases of deformation and grain displacement were defined using the stationary position of the subsidence surface: a densification-dominated phase and a convection-dominated phase. However, it should be noted that, based on our observations, the surface reaching its stationary position is a progressive process requiring many cycles.

On the contrary, the change in slope configuration is not significant in Test 3, as shown in Fig. J.5d. The experiment started with a surface inclination of $34°$. The evolution of the surface shows that the slope inclination increased and reached $35°$, and the surface dropped slightly. In addition, no global heave was formed throughout the experiment, since the material on the upper slope slipped when the slope reached the critical inclination angle of $35°$, preventing the formation of a global heave.

It is noted that Figs. J.5c and J.5d illustrate a minor influence of the cyclic tilt on the evolution of the free surface for inclined surface, when compared with Figs. J.5a and J.5b for horizontal surface. This is due to the fact that, in Tests 1 and 4, more material is placed in the depression zone, compared with Tests 2 and 4, as illustrated in Fig. J.8. As the free surface in the depression zone tends to reach the

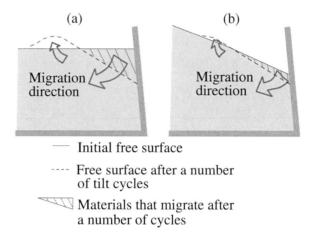

Figure J.8: Influence of the cyclic tilt on the evolution of the free surface for (a) horizontal surface and (b) inclined surface.

critical inclination angle from cycle to cycle, for the case with an initially horizontal free surface, the volume of material behind the wall that is pushed away from the wall and merges with the material behind it is larger, when compared with the cases starting with an inclined free surface. As a result, the cyclic tilt has a more significant influence for tests starting with a horizontal free surface.

In all experiments, one or two local heaves periodically appeared after the free surface reached its stationary position. After their appearances, they migrated slightly and slowly towards the wall. It was observed that a local heave temporarily disappeared when the grains forming the local heave rolled down the slope, and reappeared in the next cycles thereafter. In addition, before the material rolled down, the slope right at the local heaves, inclined by $37.5°$ to the horizontal, is larger than the critical inclination angle ($35°$). Both global and local heaves are found to associate with the appearance of passive slip surfaces. Their periodical appearances are investigated and will be presented in the next sections.

CUÉLLAR et al. [14] observed a densification of the material in the "densification-dominated phase" by evaluating the volume change of the specimen. In all our experiments, the heaved volume is also clearly less than the subsided volume, and this proves that the densification in sand also took place before the surface reached its stationary position.

J.3.2 Trajectories and eddies

Sand eddies are elliptical trajectories of a spiral or a closed loop. Their positions, shapes, circulation directions, and the time of occurrence are described in this section.

J.3.2.1 Shape and position of eddies

Trajectories obtained after the free surfaces reached their stationary positions are shown in Fig. J.9.

In these figures, eddies are elliptical in shape and are located below the subsided free surface adjacent to the wall. If we roughly describe the orbits of the particles as ellipses, then we observe that the halve axes of these ellipses do not associate with the orientations of any geometries of objects in our experiments. Closer views of the eddies, as shown in Fig. J.11, reveal that the elliptical shape is not smooth and the size of eddies varies. Moreover, eddies do not always have a closed loop. For instance, the trajectories plotted in Fig. J.11a marked by green and in Fig. J.11b marked by red have the form of an outward spiral. Such outward spiral eddies are observed in all experiments. However, inward-spiral eddies, such as those in Fig. J.11b marked by purple and blue, are also present.

The slip surface was identified by plotting the velocity field occurred at the 51th outward cycle of the wall (Fig. J.12a), along with the trajectories from cycle 50 through 100 (Fig. J.12b). The active slip surface is found to cross the eddies and the surrounding trajectories. In addition, as shown in Fig. J.13, trajectories below (or on the left side of) the active slip surface are smoother compared to those above (or on the right side of) it; and irregular zigzag trajectories are only found above the active slip surface (Fig. J.13). These large curved zigzags result from the fact that the motion of the grains located on the right side of the active slip surface is influenced by the alternating tilt directions of the wall. The zigzag patterns hence are in sync with the cyclic tilt of the wall. The curved zigzags indicate that grains on the right side of the active slip surface move generally toward the lower-right direction. The grain motion on the left side of the active slip surface is insignificant when the wall tilts outwards, leaving the accumulated displacement being mainly due to the motion occurred during inward tilts of the wall. Trajectories on this side, therefore, turn out to be smoother. These trajectories are indicating that the grains migrate to the upper left.

As a result, an active slip surface can serve as a border which distinguishes smooth from less smooth trajectories. Intersections of active slip surfaces and trajectories are defined as *"smoothness transition points*, e.g. the hollow dots marked in Fig. J.13.

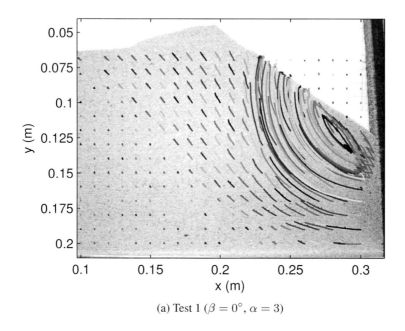

(a) Test 1 ($\beta = 0°$, $\alpha = 3$)

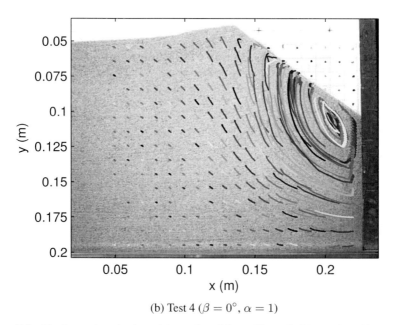

(b) Test 4 ($\beta = 0°$, $\alpha = 1$)

Figure J.9: Trajectories during (a) cycles 25 to 60 and (b) cycles 50 to 100 (To be continued). Background photographs taken at the beginning of these cycles. (*continued*)

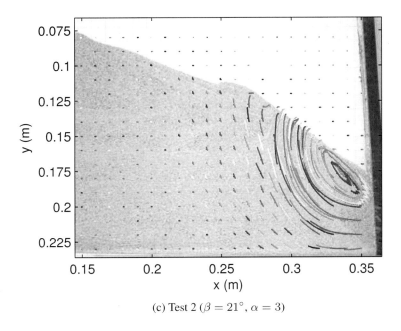

(c) Test 2 ($\beta = 21°$, $\alpha = 3$)

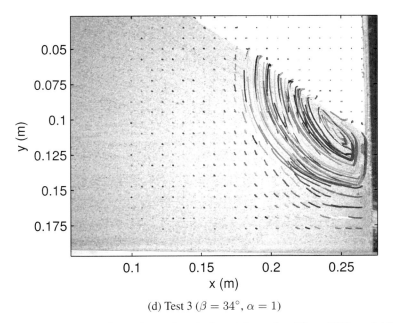

(d) Test 3 ($\beta = 34°$, $\alpha = 1$)

Figure J.10: (*continued*) Trajectories during (c) cycles 50 to 200 and (d) cycles 50 to 200. Background photographs taken at the beginning of these cycles.

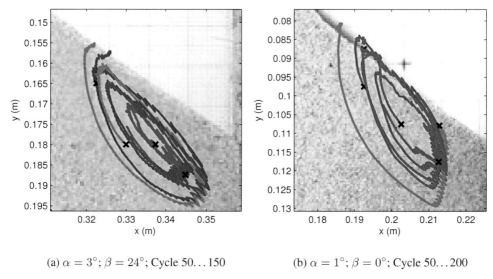

(a) $\alpha = 3°$; $\beta = 24°$; Cycle 50...150   (b) $\alpha = 1°$; $\beta = 0°$; Cycle 50...200

Figure J.11: Examples of various sand eddies in (a) Test 2 (background at cycle 50) and (b) Test 4 (Background at cycle 200). The origins of paths are marked with '×'.

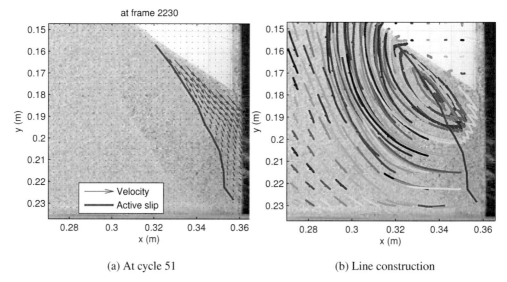

(a) At cycle 51   (b) Line construction

Figure J.12: (a) Active slip surface determined using the velocity field at cycle 51 while the wall was tilting outwards. (b) Comparison of the active slip surface and trajectories plotted during the cycles 50 to 100 in Test 2 ($\alpha = 3°, \beta = 21°$).

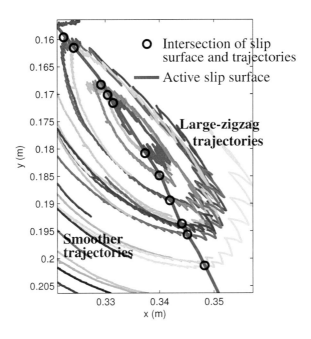

Figure J.13: Close-up view of Fig. J.12b with the background removed. The trajectories above the active slip surface are large zigzags which are less smooth. On the contrary, the trajectories below the slip surface are smoother.

Defining the smoothness transition points allows positioning of eddies in such experiments, since these points are situated on an active slip surface. If an eddy is pictured as an ellipse, the smoothness transition points are located around the antipodal points of the ellipse and the major axis of the ellipse shall approximately coincide with the active slip surface. This also explains the orientation of all eddies in Fig. J.9.

Information found in Fig. J.13 helps to discover the area in which eddies occur, as illustrated in Fig. J.14. The area, marked by green (Fig. J.14a), is enclosed by the free surface, the trajectory which is below the active slip surface passing the intersection of the free surface and the active slip surface, and a boundary enveloping the zigzag trajectories (marked by a black dashed line).

J.3.2.2 Zoning and passive slip surfaces

CUÉLLAR et al. [14] indicated that the movements of the grains possess different characteristics depending on where the grains are. They defined two domains based on the profile cutting vertically into the soil specimen along the loading direction (Fig. J.15): (a) The "convected domain" where the convective flow of grains takes

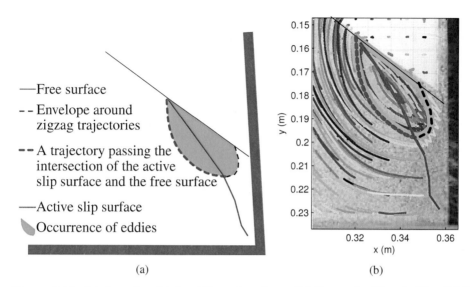

——Free surface

- - Envelope around
  zigzag trajectories

-- A trajectory passing the
   intersection of the active
   slip surface and the free surface

——Active slip surface

◗ Occurrence of eddies

(a)   (b)

Figure J.14: (a) Area in which eddies take place. (b) Example: Test 2 (Fig. J.12b).

place. (b) The "static domain" where grain displacements are less significant. Between the two domains, there exists a thin transition band.

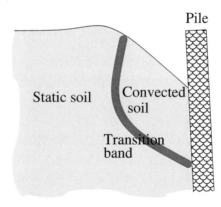

Figure J.15: Illustration of soil domains defined in [14].

Different characteristics in grain motions, by means of trajectories, are also observed in our results. Take Test 1 as an example, the zoning based on the characteristics of trajectories is shown in Fig. J.16a and three domains are defined. Domain I corresponds to the convected domain and Domain II and III correspond to the static domain defined by CUÉLLAR et al. [14] (Fig. J.15).

In domain I, eddies, large-zigzag trajectories and smoother trajectories are present. Grains on eddies will migrate elliptically until they leave the eddies, ending up with

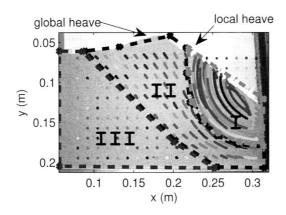

Figure J.16: Zoning by means of trajectories plotted from cycle 25 through 50 in Test 1.

re-appearing on the surface. This might explain the re-appearance of the red grains on the soil surface in [14]. In their experiment, the red grains were initially allocated on the free surface. They sank beneath the surface within the depression zone and re-appeared again on the free surface after a number of cycles. Trajectories in domain II are significantly shorter than those in domain I, suggesting that only minor displacement is accumulated, inducing the formation of the global heave. The size of domain II is associated with the initial surface inclination angle (compare plots in Fig. J.16). In particular, the smaller β is, the wider domain II will be. It is noted that domain II can be barely recognized in Test 3 (Fig. J.9) that started with a surface inclination angle of $34°$. In domain III, the dot-like trajectories indicate little to no displacement. That explains the negligible alteration of the free surface on top of this domain. The evolution of the domains in size and shape throughout the experiments as well as the formation and the migration of global heaves can be reconstructed by the information provided in section J.3.3, as passive slip surfaces are natural borders of domains.

According to our results (see section J.3.3) and the observation made by BOBRYAKOV et al. [7], the border of domains I and II migrates from cycle to cycle as will be shown in section J.3.3.

J.3.2.3   Circulation directions of eddies

Fig. J.17 qualitatively illustrates the evolution directions of the trajectories based on Fig. J.9. The evolution directions of the trajectories above the active slip surface are opposite to those below the active slip surface. As a result, the circulation directions of the eddies are clockwise (in the case that the cyclic retaining wall is on the right side). This is valid for all tests we have carried out.

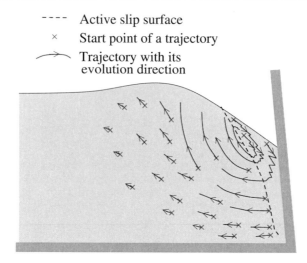

Figure J.17: Evolution directions of the trajectories.

J.3.2.4 Time of occurrence of eddies

CUÉLLAR et al. speculated that "..., it is possible that some convective grain migration already takes place simultaneously with the densification during the first loading cycles ...". This speculation is confirmed by our observations that eddies occurred in the beginning of all experiments and the positions of the eddies during an experiment are related to the evolving positions of the free surface adjacent to the wall. For instance, in Test 1 ($\beta = 0°$, $\alpha = 3°$), as shown in Fig. J.18a, during the starting cycles 1 to 11, the eddies were tracked; as well as in Fig. J.18b during the starting n 1 to 19.

J.3.3 Evolution of passive slip surfaces

The repeated process of the appearance of new passive slip surfaces and their migration towards the wall has been observed by BOBRYAKOV et al. [7]. In particular, after the appearance of a new slip surface, it shifts towards the wall and disappears at a certain position. During the course of its migration, another new passive slip surface is formed at a distance behind the existing one. Such repeated appearance of passive slip surfaces was also observed in domain I in our experiments, as shown in Figs. J.19, J.20 and J.21. In addition, these figures also suggest that with a smaller tilt amplitude ($\alpha = 1°$) (a) more cycles are required before the first passive slip surface became visible and (b) the presence of a passive slip surface lasted more cycles. It should be noted that an A-surface denotes a passive slip surface occurring in domain I (e.g. Fig. J.16c) where the significant movements of grains to the up-left takes place. A B-surface denotes a slip in domain II where minor motion occurs. In case

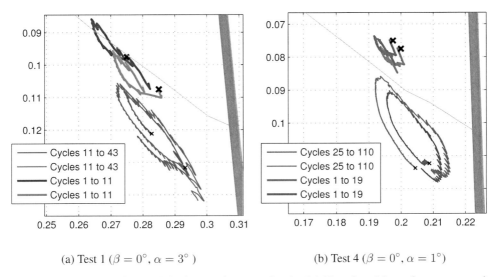

(a) Test 1 ($\beta = 0°$, $\alpha = 3°$)   (b) Test 4 ($\beta = 0°$, $\alpha = 1°$)

Figure J.18: Eddies formed during various cycles in (a) Test 1, with surface captured at cycle 13, (b) Test 4, with surface captured at cycle 107.

more than one surface is present, the sequence of occurrences of slips is stated by a number, with a greater number denoting a later appearance (e.g., A2-surface occurs later than A1-surface).

No apparent trend of evolution of B-surfaces was noticed among experiments. Each stayed more or less at the same position during its presence. It is noted that B-surfaces were identified in Test 1 ($\beta = 0°$) and in Test 4 ($\beta = 21°$), but no B-surface occurred in Test 3 ($\beta = 40°$). This might result from the fact that domain II is very thin in Test 3, as shown in Fig. J.9. In addition, B-surface was not exhibited at all times in Test 4. A missing border between domains II and III indicates that no clear slip surfaces were present, implying gradual changes of the velocity fields in direction and magnitude.

J.3.4   Evolution of void ratio

The PIV evaluation program yields the displacement vectors $\mathbf{u} = (u_1, u_2, u_3)$ at the nodes of a regular array (grid). The changes of the void ratio e are obtained from the equation

$$\dot{e} = (1 + e)\mathrm{tr}\mathbf{D} \tag{J.1}$$

or

$$\Delta e = (1 + e)\left(\frac{\partial u_1}{\partial x_1} + \frac{\partial u_2}{\partial x_2} + \frac{\partial u_3}{\partial x_3}\right) \tag{J.2}$$

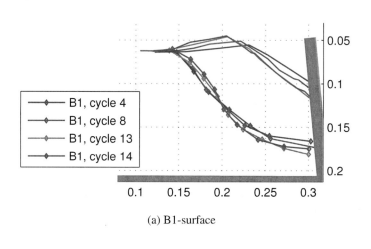

(a) B1-surface

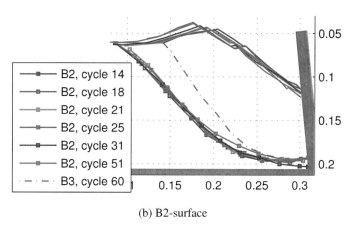

(b) B2-surface

Figure J.19: Evolution of passive slip surfaces in Test 1. A-surface and B-surface are denoted by A and B. Unit of axis in meter. (*continued*)

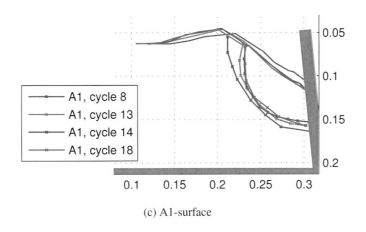

(c) A1-surface

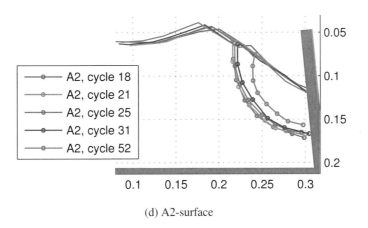

(d) A2-surface

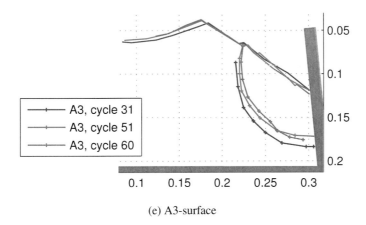

(e) A3-surface

Figure J.19: (*continued*).

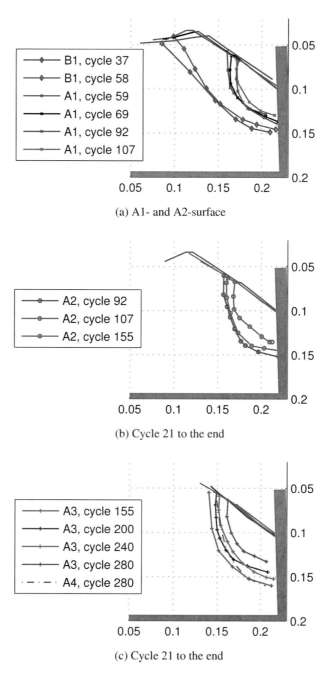

(a) A1- and A2-surface

(b) Cycle 21 to the end

(c) Cycle 21 to the end

Figure J.20: Evolution of passive slip surfaces in Test 4. A-surface and B-surface are denoted by A and B

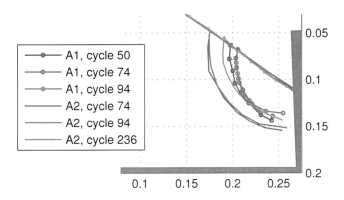

Figure J.21: Evolution of passive slip surfaces in Test 3. A-surface and B-surface are denoted by A and B.

The spatial derivatives are obtained by the linear approximation of the displacement vector field. They are affected by the number of points n used in approximation and are unrealistically large on the free surface, adjacent to the tilting bar and on the slip surfaces where the displacement vector field changes abruptly. Therefore, a threshold χ is used to cut off too large derivatives: For any integers i and j,

$$\text{if } |u_{i,j}| > \chi, \quad \text{then} \quad u_{i,j} = 0 \tag{J.3}$$

In addition, as Δe is an approximation, considerable error arises while integrating it over time. For this reason, a factor η is used to backscale it:

$$e^{t+\Delta t} = e^t + \eta \Delta e \tag{J.4}$$

Taking Test 1 as an example, the so obtained void ratios in the backfill after 10 cycles is shown Fig. J.22. Improper selections of the three parameters n, χ and η can lead to wrong estimation of Δe. For instance, Fig. J.22a shows that void ratios pertain their initial values of 0.77 over 10 cycles, contradicting to the general agreement that the backfill becomes denser [7, 14]. The other three combinations are shown in Figs. J.22b, J.22c and J.22d. These results comply with the general agreement that the backfill gets denser over the cycles.

Although the so obtained void ratios do not represent the exact ones, the tendency of their evolution provide valuable information on how density and/or pore space changes in such experiments. Fig. J.23 shows the evolution of void ratio, using $n = 49, \chi = 0.002, \eta = 0.7$, in cycles 5, 10, 15, 25, 35, and 45.

These figures indicate that the backfill tends to become denser except of the passive slip surfaces, where shear bands appear (see Fig. J.19 for comparisons). With

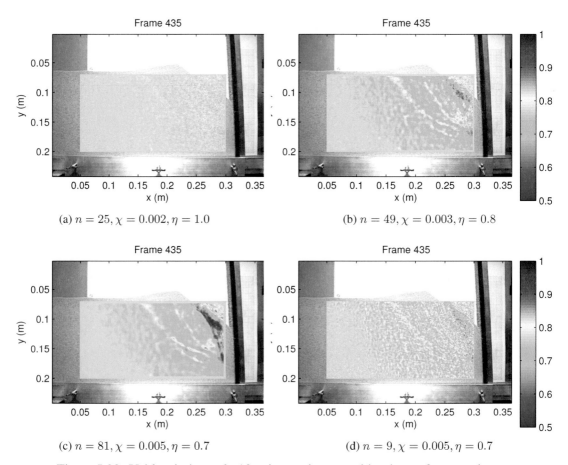

(a) $n = 25, \chi = 0.002, \eta = 1.0$

(b) $n = 49, \chi = 0.003, \eta = 0.8$

(c) $n = 81, \chi = 0.005, \eta = 0.7$

(d) $n = 9, \chi = 0.005, \eta = 0.7$

Figure J.22: Void ratio in cycle 10 using various combinations of n, χ and η.

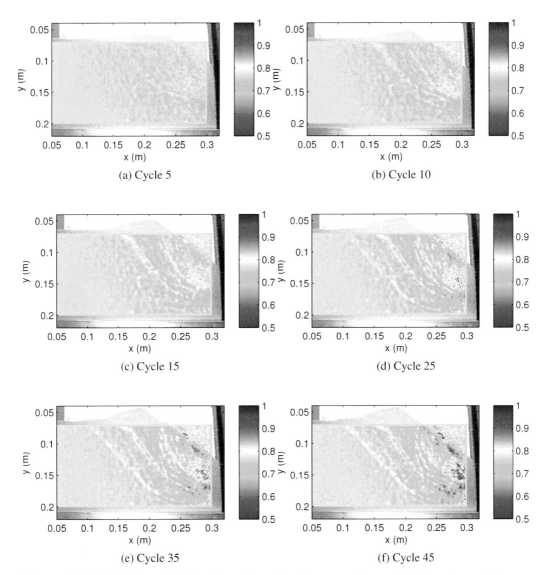

Figure J.23: Evolution of void ratio evaluated using $n = 49, \chi = 0.002$ and $\eta = 0.7$ in Test 1, with initial void radio $e = 0.77$. (*continued*)

increasing number of cycles, the density on the bottom of zone I decreases. This might result from the fact that the migration of the passive slip surfaces occurs in zone I.

J.4   Conclusions

The combination of the model test in plane strain conditions and the PIV technique makes it possible to capture the grain motion subject to cyclic tilt of a wall. Among all the findings presented in the previous sections, the most remarkable ones are: (a) The visualization of the grain motion in terms of trajectories and velocity fields. (b) The understanding of the occurrence of eddies, namely the grains on the right side of the active slip surface exhibit the zigzag pattern in response to the cyclic tilt, moving to the lower right toward the wall until crossing the active slip surface to the other side, whereas those on the left side migrate to the upper left, end up with re-appearing on the surface or crossing the active slip surface. As a result of these two mechanisms, either closed or spiral non-smooth trajectories, i.e. sand eddies, occurred around the active slip surface. (c) Experiments with various tilt amplitudes and free surface configurations were carried out to study those effects. The width of domain II is found to be inversely proportional to the initial surface inclination angle, more cycles are required to form eddies of the same size if the tilt amplitude is smaller. (d) The evolution of void ratio was visualized. Besides all these findings, the experimental results themselves (e.g. the re-appearance migration of the passive slip surfaces) provide particle-based numerical simulation methods with challenging examples where a granular material undergoes large deformation and sophisticated mixing processes.

Acknowledgements
The authors would like to thank the students Amy Cleaves and Eleanor Shirley for carrying out some of the experiments presented in herein.

Appendix K

One-dimensional Example

A simple one-dimensional example is given herein to demonstrate the following:

1. The framework of SPARC: The approximation of spatial derivatives, calculation of governing equations, and time advance. (For the demonstration purpose of the hand-calculation, Newton's method, substeps and Runge-Kutta method are not used.)

2. The arc-length method. It is to demonstrate how to use the arc-length method to solve this problem.

3. The pseudo-arc-length method. It is to demonstrate how to use the pseudo-arc-length method to solve this problem.

The boundary condition of a one-dimensional oedometric axial extension test is shown in Fig. K.1. The system consists of three particles, p_1, p_2, and p_3.

K.1 Solution to the example using SPARC

Recall that the solution procedure of SPARC follows:
Velocity field ($\mathbf{v}^t$, unknowns) $\rightsquigarrow$ Neighbor search $\rightsquigarrow$ Velocity gradient $\mathbf{L}^t = \nabla^t \mathbf{v}^t$ $\rightsquigarrow$ Stretching and spin tensors $\mathbf{D}^t, \mathbf{W}^t \rightsquigarrow$ Stress rate tensor $\overset{\circ}{\mathbf{T}}{}^t$ (constitutive model) $\rightsquigarrow \dot{\mathbf{T}}^t = \overset{\circ}{\mathbf{T}}{}^t - \mathbf{T}^t \mathbf{W}^t + \mathbf{W}^t \mathbf{T}^t$ (JAUMANN-ZARENBA) $\rightsquigarrow$ Stress tensor $\mathbf{T}^{t+\Delta t} = \mathbf{T}^t + \dot{\mathbf{T}}^t \Delta t \rightsquigarrow$ Solve $\nabla^{t+\Delta t} \cdot \mathbf{T}^{t+\Delta t} = \mathbf{0}$.

Velocity field and position vector of the problem

$$\mathbf{v}^t_{(p_1)} = \begin{bmatrix} 0 & 0 & 0 \end{bmatrix} \qquad \mathbf{x}^t_{(p_1)} = \begin{bmatrix} 0 & 0 & 0 \end{bmatrix}$$
$$\mathbf{v}^t_{(p_2)} = \begin{bmatrix} 0 & 0 & v^t_{3(p_2)} \end{bmatrix} \qquad \mathbf{x}^t_{(p_2)} = \begin{bmatrix} 0 & 0 & 1 \end{bmatrix}$$
$$\mathbf{v}^t_{(p_3)} = \begin{bmatrix} 0 & 0 & 1 \end{bmatrix} \qquad \mathbf{x}^t_{(p_3)} = \begin{bmatrix} 0 & 0 & 2 \end{bmatrix}$$

Given:

1. The spacing between any two adjacent particles is 1 m.
2. The velocities of the particles are $v_{3(p_1)} = 0$ m/s and $v_{3(p_3)} = 1$ m/s.
3. The initial stress states of all points are

$$\mathbf{T}^t = \begin{bmatrix} 0 & 0 & 0 \\ 0 & 0 & 0 \\ 0 & 0 & 0 \end{bmatrix}$$

4. Material model: Elasticity with Young's modulus $E = 2$ GPa, Possion's ratio $\nu = 0.25$. Gravity is neglected, i.e. $\mathbf{g} = 0$

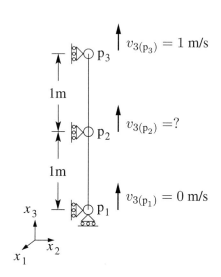

Figure K.1: Boundary conditions.

Task:
Determination of $\mathbf{v}_{(p_2)}$ and the stress state of all particles at $t + \Delta t$, with $\Delta t = 0.01$ sec.

Neighbors

Here we use fixed-radius neighbor search with radius of support $r = 1.5$ m. Consequently, the neighbors of particles are as follows:

Neighbors of p_1 : p_1 , p_2
Neighbors of p_2 : p_1, p_2, p_3
Neighbors of p_3 : p_2 , p_3

Calculation of velocity gradient $\mathbf{L}^t = \nabla^t \mathbf{v}^t$

Since the problem is one-dimensional, $\mathbf{L}^t = \begin{bmatrix} 0 & 0 & 0 \\ 0 & 0 & 0 \\ 0 & 0 & L_{33}^t \end{bmatrix}$. Consequently, only the third components of the velocity and position vectors are needed in the following calculations.

For particle p_1 (with neighbors p_1 and p_2):

$$\bar{\boldsymbol{x}}^t_{(p_1)} = \begin{bmatrix} x^t_{3(p_1)} \\ x^t_{3(p_2)} \end{bmatrix} - \begin{bmatrix} x^t_{3(p_1)} \\ x^t_{3(p_1)} \end{bmatrix} = \begin{bmatrix} 0 \\ 1 \end{bmatrix} - \begin{bmatrix} 0 \\ 0 \end{bmatrix} = \begin{bmatrix} 0 \\ 1 \end{bmatrix} \tag{K.1}$$

$$\bar{\boldsymbol{\mathcal{V}}}^t_{(p_1)} = \begin{bmatrix} v^t_{3(p_1)} \\ v^t_{3(p_2)} \end{bmatrix} - \begin{bmatrix} v^t_{3(p_1)} \\ v^t_{3(p_1)} \end{bmatrix} = \begin{bmatrix} 0 \\ v^t_{3(p_2)} \end{bmatrix} - \begin{bmatrix} 0 \\ 0 \end{bmatrix} = \begin{bmatrix} 0 \\ v^t_{3(p_2)} \end{bmatrix} \tag{K.2}$$

$$\rightsquigarrow \qquad L^t_{33(p_1)} = \left[\bar{\boldsymbol{\mathcal{X}}}^t_{(p_1)} \right]^{-1} \bar{\boldsymbol{\mathcal{V}}}^t_{(p_1)} = \begin{bmatrix} 0 \\ 1 \end{bmatrix}^{-1} \begin{bmatrix} 0 \\ v^t_{3(p_2)} \end{bmatrix} = \begin{bmatrix} 0 & 1 \end{bmatrix} \begin{bmatrix} 0 \\ v^t_{3(p_2)} \end{bmatrix} = v^t_{3(p_2)} \tag{K.3}$$

$$\text{or} \qquad L^t_{33(p_1)} = \frac{\bar{\mathcal{V}}^t_{2(p_1)} - \bar{\mathcal{V}}^t_{1(p_1)}}{\bar{\mathcal{X}}^t_{2(p_1)} - \bar{\mathcal{X}}^t_{1(p_1)}} = \frac{v^t_{3(p_2)} - 0}{1 - 0} = v^t_{3(p_2)} \tag{K.4}$$

For particle p_2 (with neighbors p_1, p_2 and p_3):

$$\bar{\boldsymbol{\mathcal{X}}}^t_{(p_2)} = \begin{bmatrix} x^t_{3(p_1)} \\ x^t_{3(p_2)} \\ x^t_{3(p_3)} \end{bmatrix} - \begin{bmatrix} x^t_{3(p_2)} \\ x^t_{3(p_2)} \\ x^t_{3(p_2)} \end{bmatrix} = \begin{bmatrix} 0 \\ 1 \\ 2 \end{bmatrix} - \begin{bmatrix} 1 \\ 1 \\ 1 \end{bmatrix} = \begin{bmatrix} -1 \\ 0 \\ 1 \end{bmatrix} \tag{K.5}$$

$$\bar{\boldsymbol{\mathcal{V}}}^t_{(p_2)} = \begin{bmatrix} v^t_{3(p_1)} \\ v^t_{3(p_2)} \\ v^t_{3(p_3)} \end{bmatrix} - \begin{bmatrix} v^t_{3(p_2)} \\ v^t_{3(p_2)} \\ v^t_{3(p_2)} \end{bmatrix} = \begin{bmatrix} 0 \\ v^t_{3(p_2)} \\ 1 \end{bmatrix} - \begin{bmatrix} v^t_{3(p_2)} \\ v^t_{3(p_2)} \\ v^t_{3(p_2)} \end{bmatrix} = \begin{bmatrix} -v^t_{3(p_2)} \\ 0 \\ 1 - v^t_{3(p_2)} \end{bmatrix} \tag{K.6}$$

$$\rightsquigarrow \qquad L^t_{33(p_2)} = \left[\bar{\boldsymbol{\mathcal{X}}}^t_{(p_2)} \right]^{-1} \bar{\boldsymbol{\mathcal{V}}}^t_{(p_2)} = \begin{bmatrix} -1 \\ 0 \\ 1 \end{bmatrix}^{-1} \begin{bmatrix} -v^t_{3(p_2)} \\ 0 \\ 1 - v^t_{3(p_2)} \end{bmatrix} \tag{K.7}$$

$$= \begin{bmatrix} -0.5 & 0 & 0.5 \end{bmatrix} \begin{bmatrix} -v^t_{3(p_2)} \\ 0 \\ 1 - v^t_{3(p_2)} \end{bmatrix} = 0.5 \tag{K.8}$$

$$\text{or} \qquad L^t_{33(p_2)} = \frac{\mathcal{V}^t_{3(p_2)} - \mathcal{V}^t_{1(p_2)}}{\mathcal{X}^t_{3(p_2)} - \mathcal{X}^t_{1(p_2)}} = \frac{(1 - v^t_{3(p_2)}) - (-v^t_{3(p_2)})}{1 - (-1)} = 0.5 \tag{K.9}$$

For particle p_3 (with neighbors p_2 and p_3):

$$\bar{\boldsymbol{\mathcal{X}}}^t_{(p_3)} = \begin{bmatrix} x^t_{3(p_2)} \\ x^t_{3(p_3)} \end{bmatrix} - \begin{bmatrix} x^t_{3(p_3)} \\ x^t_{3(p_3)} \end{bmatrix} = \begin{bmatrix} 1 \\ 2 \end{bmatrix} - \begin{bmatrix} 2 \\ 2 \end{bmatrix} = \begin{bmatrix} -1 \\ 0 \end{bmatrix} \tag{K.10}$$

$$\bar{\boldsymbol{\mathcal{V}}}^t_{(p_3)} = \begin{bmatrix} v^t_{3(p_2)} \\ v^t_{3(p_3)} \end{bmatrix} - \begin{bmatrix} v^t_{3(p_3)} \\ v^t_{3(p_3)} \end{bmatrix} = \begin{bmatrix} v^t_{3(p_2)} \\ 1 \end{bmatrix} - \begin{bmatrix} 1 \\ 1 \end{bmatrix} = \begin{bmatrix} v^t_{3(p_2)} - 1 \\ 0 \end{bmatrix} \tag{K.11}$$

$$\rightsquigarrow \qquad L^t_{33(p_3)} = \left[\bar{\boldsymbol{\mathcal{X}}}^t_{(p_3)} \right]^{-1} \bar{\boldsymbol{\mathcal{V}}}^t_{(p_3)} = \begin{bmatrix} -1 \\ 0 \end{bmatrix}^{-1} \begin{bmatrix} v^t_{3(p_2)} - 1 \\ 0 \end{bmatrix} \tag{K.12}$$

$$= \begin{bmatrix} -1 & 0 \end{bmatrix} \begin{bmatrix} v_{3(p_2)}^t - 1 \\ 0 \end{bmatrix} = 1 - v_{3(p_2)}^t \qquad \text{(K.13)}$$

$$\text{or} \qquad L_{33(p_3)}^t = \frac{\mathcal{V}_{3(p_3)}^t - \mathcal{V}_{2(p_3)}^t}{\mathcal{X}_{3(p_3)}^t - \mathcal{X}_{2(p_3)}^t} = \frac{0 - (v_{3(p_2)}^t - 1)}{0 - (-1)} = 1 - v_{3(p_2)}^t \qquad \text{(K.14)}$$

Now, we have obtained the velocity gradients of the particles:

$$\mathbf{L}_{(p_1)}^t = \begin{bmatrix} 0 & 0 & 0 \\ 0 & 0 & 0 \\ 0 & 0 & v_{3(p_2)}^t \end{bmatrix}, \quad \mathbf{L}_{(p_2)}^t = \begin{bmatrix} 0 & 0 & 0 \\ 0 & 0 & 0 \\ 0 & 0 & 0.5 \end{bmatrix}, \quad \mathbf{L}_{(p_3)}^t = \begin{bmatrix} 0 & 0 & 0 \\ 0 & 0 & 0 \\ 0 & 0 & 1 - v_{3(p_2)}^t \end{bmatrix}$$

Calculation of stretching and spin tensor (D, W)

Recall $\mathbf{D} = \dfrac{\mathbf{L} + \mathbf{L}^{\mathrm{T}}}{2}$ and $\mathbf{W} = \dfrac{\mathbf{L} - \mathbf{L}^{\mathrm{T}}}{2}$, we have

$$\mathbf{D}_{(p_1)}^t = \begin{bmatrix} 0 & 0 & 0 \\ 0 & 0 & 0 \\ 0 & 0 & v_{3(p_2)}^t \end{bmatrix}, \mathbf{D}_{(p_2)}^t = \begin{bmatrix} 0 & 0 & 0 \\ 0 & 0 & 0 \\ 0 & 0 & 0.5 \end{bmatrix}, \mathbf{D}_{(p_3)}^t = \begin{bmatrix} 0 & 0 & 0 \\ 0 & 0 & 0 \\ 0 & 0 & 1 - v_{3(p_2)}^t \end{bmatrix}$$
$$\text{(K.15)}$$

$$\mathbf{W}_{(p_1)}^t = \mathbf{W}_{(p_2)}^t = \mathbf{W}_{(p_3)}^t = \begin{bmatrix} 0 & 0 & 0 \\ 0 & 0 & 0 \\ 0 & 0 & 0 \end{bmatrix} \qquad \text{(K.16)}$$

Calculation of stress rate tensor $\overset{\circ}{\mathbf{T}}{}^t$ and stress tensor $\mathbf{T}^{t+\Delta t}$

Given $E = 2$ GPa and $\nu = 0.25$, the constitutive model based on the Hooke's law reads:

$$\begin{bmatrix} \overset{\circ}{\mathbf{T}}{}_{11}^t \\ \overset{\circ}{\mathbf{T}}{}_{22}^t \\ \overset{\circ}{\mathbf{T}}{}_{33}^t \\ \overset{\circ}{\mathbf{T}}{}_{23}^t \\ \overset{\circ}{\mathbf{T}}{}_{31}^t \\ \overset{\circ}{\mathbf{T}}{}_{12}^t \end{bmatrix} = \frac{E}{(1 + v)(1 - 2v)} \cdot$$

$$
\begin{bmatrix}
(1-v) & v & v & 0 & 0 & 0 \\
v & (1-v) & v & 0 & 0 & 0 \\
v & v & (1-v) & 0 & 0 & 0 \\
0 & 0 & 0 & \frac{(1-2v)}{2} & 0 & 0 \\
0 & 0 & 0 & 0 & \frac{(1-2v)}{2} & 0 \\
0 & 0 & 0 & 0 & 0 & \frac{(1-2v)}{2}
\end{bmatrix}
\begin{bmatrix}
D_{11}^t \\
D_{22}^t \\
D_{33}^t \\
2D_{23}^t \\
2D_{31}^t \\
2D_{12}^t
\end{bmatrix}
\tag{K.17}
$$

$$
= 10^9 \cdot
\begin{bmatrix}
2.4 & 0.8 & 0.8 & 0 & 0 & 0 \\
0.8 & 2.4 & 0.8 & 0 & 0 & 0 \\
0.8 & 0.8 & 2.4 & 0 & 0 & 0 \\
0 & 0 & 0 & 0.8 & 0 & 0 \\
0 & 0 & 0 & 0 & 0.8 & 0 \\
0 & 0 & 0 & 0 & 0 & 0.8
\end{bmatrix}
\begin{bmatrix}
D_{11}^t \\
D_{22}^t \\
D_{33}^t \\
2D_{23}^t \\
2D_{31}^t \\
2D_{12}^t
\end{bmatrix}
\tag{K.18}
$$

With $\dot{\mathbf{T}}^t = \overset{\circ}{\mathbf{T}}{}^t - \mathbf{T}^t \mathbf{W}^t + \mathbf{W}^t \mathbf{T}^t$, the stress tensors $\mathbf{T}^{t+\Delta t} = \mathbf{T}^t + \dot{\mathbf{T}}^t \Delta t$ for particles p_1, p_2 and p_3 are obtained:

$$
\mathbf{T}_{(p_1)}^{t+\Delta t} = 10^9 \cdot
\begin{bmatrix}
0.8 & 0 & 0 \\
0 & 0.8 & 0 \\
0 & 0 & 2.4v_{3(p_2)}^t
\end{bmatrix}
\Delta t
\tag{K.19}
$$

$$
\mathbf{T}_{(p_2)}^{t+\Delta t} = 10^9 \cdot
\begin{bmatrix}
0.4 & 0 & 0 \\
0 & 0.4 & 0 \\
0 & 0 & 2.4 \cdot 0.5
\end{bmatrix}
\Delta t
\tag{K.20}
$$

$$
\mathbf{T}_{(p_3)}^{t+\Delta t} = 10^9 \cdot
\begin{bmatrix}
0.8 & 0 & 0 \\
0 & 0.8 & 0 \\
0 & 0 & 2.4(1 - v_{3(p_2)}^t)
\end{bmatrix}
\Delta t
\tag{K.21}
$$

Calculation of the system of equations

The stress tensor $\mathbf{T}^{t+\Delta t}$ must be calculated herein only for computing the function $\mathbf{y} = \nabla^{t+\Delta t} \cdot \mathbf{T}^{t+\Delta t}$ of a particle. This function is the left-hand side (residuals) of the governing equation $\nabla^{t+\Delta t} \cdot \mathbf{T}^{t+\Delta t} + \rho \mathbf{g} = \mathbf{0}$ (with $\mathbf{g} = \mathbf{0}$). The function $\mathbf{y}$ consists of three components, $y_i = \partial T_{ji}^{t+\Delta t}/\partial x_j$ ($i = 1 \cdots 3$) which correspond to the three components of the velocity vector, v_i^t ($i = 1 \cdots 3$). If a component v_i^t of a particle is an unknown, the corresponding y_i at that particle needs to be calculated, and $y_i = 0$ must be solved for the determination of v_i^t. Thus, for this problem, only y_3 for particle p_2 (denoted as $y_{3(p_2)}$) is evaluated, and the equation $y_{3(p_2)} = 0$ will be solved.

Recall that $y_3 = \partial T_{j3}^{t+\Delta t}/\partial x_j = \partial T_{13}^{t+\Delta t}/\partial x_1 + \partial T_{23}^{t+\Delta t}/\partial x_2 + \partial T_{33}^{t+\Delta t}/\partial x_3$. In this one-dimensional problem, $\partial T_{13}^{t+\Delta t}/\partial x_1 = 0$ and $\partial T_{23}^{t+\Delta t}/\partial x_2 = 0$. Thus, only the components $T_{33}^{t+\Delta t}$ of all particles are needed in the following calculation of the spatial derivative $\partial T_{33}^{t+\Delta t}/\partial x_3$ at particle p_2:

$$\boldsymbol{\mathcal{T}}_{(p_2)} = \begin{bmatrix} T_{33(p_1)}^{t+\Delta t} \\ T_{33(p_2)}^{t+\Delta t} \\ T_{33(p_3)}^{t+\Delta t} \end{bmatrix} - \begin{bmatrix} T_{33(p_2)}^{t+\Delta t} \\ T_{33(p_2)}^{t+\Delta t} \\ T_{33(p_2)}^{t+\Delta t} \end{bmatrix} = \left(\begin{bmatrix} v_{3(p_2)}^t \\ 0.5 \\ (1 - v_{3(p_2)}^t) \end{bmatrix} - \begin{bmatrix} 0.5 \\ 0.5 \\ 0.5 \end{bmatrix} \right) \cdot 2.4 \cdot 10^9 \Delta t$$

(K.22)

$$= \begin{bmatrix} v_{3(p_2)}^t - 0.5 \\ 0 \\ 0.5 - v_{3(p_2)}^t \end{bmatrix} \cdot 2.4 \cdot 10^9 \Delta t$$

(K.23)

As the equilibrium is fulfilled at $t + \Delta t$, the position must be updated with $\mathbf{x}^{t+\Delta t} = \mathbf{x}^t + \mathbf{v}^t \Delta t$:

$$x_{3(p_1)}^{t+\Delta t} = x_{3(p_1)}^t + v_{3(p_1)}^t \Delta t = 0$$
$$x_{3(p_2)}^{t+\Delta t} = x_{3(p_2)}^t + v_{3(p_2)}^t \Delta t = 1 + v_{3(p_2)}^t \Delta t$$
$$x_{3(p_3)}^{t+\Delta t} = x_{3(p_3)}^t + v_{3(p_3)}^t \Delta t = 2 + 1 \cdot \Delta t$$

Thus, we have

$$\boldsymbol{\mathcal{X}}_{(p_2)}^{t+\Delta t} = \begin{bmatrix} x_{3(p_1)}^{t+\Delta t} \\ x_{3(p_2)}^{t+\Delta t} \\ x_{3(p_3)}^{t+\Delta t} \end{bmatrix} - \begin{bmatrix} x_{3(p_2)}^{t+\Delta t} \\ x_{3(p_2)}^{t+\Delta t} \\ x_{3(p_2)}^{t+\Delta t} \end{bmatrix} = \begin{bmatrix} -(1 + v_{3(p_2)}^t \Delta t) \\ 0 \\ (2 + \Delta t) - (1 + v_{3(p_2)}^t \Delta t) \end{bmatrix}$$

(K.24)

and the spatial derivative

$$\frac{\partial T_{33(p_3)}^{t+\Delta t}}{\partial x_3} = [\boldsymbol{\mathcal{X}}_{(p_2)}^{t+\Delta t}]^{-1} \boldsymbol{\mathcal{T}}_{(p_2)}^{t+\Delta t} = \frac{T_{1(p_2)}^{t+\Delta t} - T_{3(p_2)}^{t+\Delta t}}{\boldsymbol{\mathcal{X}}_{1(p_2)}^{t+\Delta t} - \boldsymbol{\mathcal{X}}_{3(p_2)}^{t+\Delta t}}$$

(K.25)

$$= \frac{(2v_{3(p_2)}^t - 1) \cdot 2.4 \cdot 10^9 \Delta t}{2 + \Delta t}$$

(K.26)

The velocity is then determined by solving the equation $y_{3(p_2)} = \frac{\partial T_{j3(p_2)}^{t+\Delta t}}{\partial x_j} = 0$

$$\frac{\partial T_{j3(p_2)}^{t+\Delta t}}{\partial x_j} = \frac{\partial T_{13(p_2)}^{t+\Delta t}}{\partial x_1} + \frac{\partial T_{23(p_2)}^{t+\Delta t}}{\partial x_2} + \frac{\partial T_{33(p_2)}^{t+\Delta t}}{\partial x_3} = 0$$

$$\rightsquigarrow \quad 0 + 0 + \frac{(2v_{3(p_2)}^t - 1) \cdot 2.4 \cdot 10^9 \Delta t}{2 + \Delta t} = 0$$

$$\rightsquigarrow \quad v_{3(\mathrm{p}_2)}^t = 0.5$$

As a result, $\mathbf{v}_{(\mathrm{p}_2)}^t = \begin{bmatrix} 0 & 0 & 0.5 \end{bmatrix}$ m/s. Note that, for this problem, the system of equations contains only one equation $y_{3(\mathrm{p}_2)} = 0$.

Time advance

The time advance of position vectors for all particles:
Let

$$\mathbb{X}^t = \begin{bmatrix} \mathbf{x}_{(\mathrm{p}_1)}^t \\ \mathbf{x}_{(\mathrm{p}_2)}^t \\ \mathbf{x}_{(\mathrm{p}_3)}^t \end{bmatrix} = \begin{bmatrix} 0 & 0 & 0 \\ 0 & 0 & 1 \\ 0 & 0 & 2 \end{bmatrix}, \text{ and } \quad \mathbb{V}^t = \begin{bmatrix} \mathbf{v}_{(\mathrm{p}_1)}^t \\ \mathbf{v}_{(\mathrm{p}_2)}^t \\ \mathbf{v}_{(\mathrm{p}_3)}^t \end{bmatrix} = \begin{bmatrix} 0 & 0 & 0 \\ 0 & 0 & 0.5 \\ 0 & 0 & 1 \end{bmatrix}$$

Given $\Delta t = 0.01$, we have

$$\mathbb{X}^{t+\Delta t} = \mathbb{X}^t + \mathbb{V}^t \cdot \Delta t = \begin{bmatrix} 0 & 0 & 0 \\ 0 & 0 & 1.005 \\ 0 & 0 & 2.01 \end{bmatrix}$$

The time advance of stress: Recall that, given a velocity field, the calculation of the stress tensor $\mathbf{T}^{t+\Delta t}$ follows:
Velocity field ($\mathbf{v}^t$, unknowns) $\rightsquigarrow$ Neighbor search $\rightsquigarrow$ $\mathbf{L}^t = \nabla^t \mathbf{v}^t$ $\rightsquigarrow$ Stretching and spin tensors $\mathbf{D}^t, \mathbf{W}^t$ $\rightsquigarrow$ Stress rate tensor $\overset{\circ}{\mathbf{T}}^t$ (constitutive model) $\rightsquigarrow$ $\dot{\mathbf{T}}^t = \overset{\circ}{\mathbf{T}}^t - \mathbf{T}^t \mathbf{W}^t + \mathbf{W}^t \mathbf{T}^t$ (JAUMANN-ZARENBA) $\rightsquigarrow$ Stress tensor $\mathbf{T}^{t+\Delta t} = \mathbf{T}^t + \dot{\mathbf{T}}^t \Delta t$.
However, in order to compute residuals of the system of equation, we have obtained stress $\mathbf{T}^{t+\Delta t}$ for all particles (see eqs. K.19 through K.21). Thus, we have

$$\mathbf{T}_{(\mathrm{p}_1)}^{t+\Delta t} = 10^9 \cdot \begin{bmatrix} 0.8 & 0 & 0 \\ 0 & 0.8 & 0 \\ 0 & 0 & 2.4 v_{3(\mathrm{p}_2)}^t \end{bmatrix} \Delta t = 10^7 \cdot \begin{bmatrix} 0.8 & 0 & 0 \\ 0 & 0.8 & 0 \\ 0 & 0 & 1.2 \end{bmatrix} \tag{K.27}$$

$$\mathbf{T}_{(\mathrm{p}_2)}^{t+\Delta t} = 10^9 \cdot \begin{bmatrix} 0.4 & 0 & 0 \\ 0 & 0.4 & 0 \\ 0 & 0 & 2.4 \cdot 0.5 \end{bmatrix} \Delta t = 10^7 \cdot \begin{bmatrix} 0.8 & 0 & 0 \\ 0 & 0.8 & 0 \\ 0 & 0 & 1.2 \end{bmatrix} \tag{K.28}$$

$$\mathbf{T}_{(\mathrm{p}_3)}^{t+\Delta t} = 10^9 \cdot \begin{bmatrix} 0.8 & 0 & 0 \\ 0 & 0.8 & 0 \\ 0 & 0 & 2.4(1 - v_{3(\mathrm{p}_2)}^t) \end{bmatrix} \Delta t = 10^7 \cdot \begin{bmatrix} 0.8 & 0 & 0 \\ 0 & 0.8 & 0 \\ 0 & 0 & 1.2 \end{bmatrix} \tag{K.29}$$

K.2   Arc-length method

In this section, it is to demonstrate how the arc-length method is applied to solve this problem. For the details of the arc-length method, see Section 2.7.4 (page 27).

Recall that in the system without arc-length parameterization, only $v^t_{3(p_2)}$ (the third velocity component of p_2) is unknown $\rightsquigarrow$ Equilibrium equation is

$$y_{3(p_2)} = y_{3(p_2)}\left(v^t_{3(p_2)}\right) = 0.$$

Now, we apply arc-length method (see eq. 2.87, page 30). For that, we choose $v^t_{3(p_3)}$ as the extra parameter because the system changes with respective to the change of $v^t_{3(p_3)}$. Since $v^t_{3(p_3)}$ becomes an extra unknown (v^t_e, see eq. 2.84), the governing equation becomes (compared to eq. 2.87)

$$\widetilde{y}_3 = \widetilde{y}_3(\boldsymbol{\nu}) = \widetilde{y}_3(\underbrace{v^t_{3(p_2)}}_{\mathbf{u}^t}, \underbrace{v^t_{3(p_3)}}_{v^t_e}) = 0 \tag{K.30}$$

The arc-length parameterization provides the arc-length condition as an extra equation (eq. 2.85):

$$\sqrt{\left(v^t_{3(p_2)}\right)^2 + \left(v^t_{3(p_3)}\right)^2} = 1 \tag{K.31}$$

The new system of equation after arc-length parameterization can be illustrated in a two-dimensional diagram, as shown in Fig. K.2. The solution path is the equilibrium equation in this problem. The intersection of the solution path and the arc-length condition is the solution.

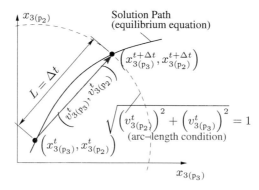

Figure K.2: Arc-length method, applied to the example. L is the arc length.

K.2.1   Solution using arc-length method: Calculation

After applying arc-length method, the new system of equations (see eq. 2.87 for comparison) consists of eqs. (K.30) and (K.31):

$$
\begin{cases}
\widetilde{y}_{3(p_2)} = \nabla^{t+\Delta t} \cdot \mathbf{T}^{t+\Delta t}_{(p_2)} = \widetilde{y}_{3(p_2)}(v^t_{3(p_2)}, v^t_{3(p_3)}) = 0 \quad \text{(equilibrium equation)} \\[2mm]
\sqrt{\left(v^t_{3(p_2)}\right)^2 + \left(v^t_{3(p_3)}\right)^2} = 1 \quad \text{(arc-length condition)}
\end{cases}
$$

Calculation of $\widetilde{y}_{3(p2)}$ and solution to the system of equations

Now the velocity component $v^t_{3(p_3)}$ becomes an unknown. Follow the solution procedure analog to eqs. (K.1) through (K.14), we obtain the velocity gradients

$$
\mathbf{L}^t_{(p_1)} = \begin{bmatrix} 0 & 0 & 0 \\ 0 & 0 & 0 \\ 0 & 0 & v^t_{3(p_2)} \end{bmatrix}, \quad
\mathbf{L}^t_{(p_2)} = \begin{bmatrix} 0 & 0 & 0 \\ 0 & 0 & 0 \\ 0 & 0 & 0.5 v^t_{3(p_3)} \end{bmatrix},
$$

$$
\mathbf{L}^t_{(p_3)} = \begin{bmatrix} 0 & 0 & 0 \\ 0 & 0 & 0 \\ 0 & 0 & v^t_{3(p_3)} - v^t_{3(p_2)} \end{bmatrix}
$$

By carrying out the calculation procedure analog to eqs. (K.15) through (K.21), the obtained stress tensors at the three particles read

$$
\mathbf{T}^{t+\Delta t}_{(p_1)} = 10^9 \cdot \begin{bmatrix} 0.8 & 0 & 0 \\ 0 & 0.8 & 0 \\ 0 & 0 & 2.4 v^t_{3(p_2)} \end{bmatrix} \Delta t
$$

$$
\mathbf{T}^{t+\Delta t}_{(p_2)} = 10^9 \cdot \begin{bmatrix} 0.4 & 0 & 0 \\ 0 & 0.4 & 0 \\ 0 & 0 & 2.4 \cdot 0.5 v^t_{3(p_3)} \end{bmatrix} \Delta t
$$

$$
\mathbf{T}^{t+\Delta t}_{(p_3)} = 10^9 \cdot \begin{bmatrix} 0.8 & 0 & 0 \\ 0 & 0.8 & 0 \\ 0 & 0 & 2.4 \left(v^t_{3(p_3)} - v^t_{3(p_2)}\right) \end{bmatrix} \Delta t
$$

Given velocities field, the position vectors at $t + \Delta t$ are

$$
x^{t+\Delta t}_{3(p_1)} = x^t_{3(p_1)} + v^t_{3(p_1)} \Delta t = 0
$$

$$
x^{t+\Delta t}_{3(p_2)} = x^t_{3(p_2)} + v^t_{3(p_2)} \Delta t = 1 + v^t_{3(p_2)} \Delta t
$$

$$
x^{t+\Delta t}_{3(p_3)} = x^t_{3(p_3)} + v^t_{3(p_3)} \Delta t - 2 + v^t_{3(p_3)} \Delta t
$$

Follow the calculation procedure analog to eqs. (K.22) through (K.26), we obtain the function $\widetilde{y}_{3(p_2)}$

$$\widetilde{y}_{3(p_2)} = \frac{\partial T^{t+\Delta t}_{j3(p_2)}}{\partial x_j} = 0 + 0 + 2.4 \cdot 10^9 \cdot \Delta t \frac{v^t_{3(p_3)} - 2v^t_{3(p_2)}}{2 + v^t_{3(p_3)}\Delta t}$$

which is a function of $v^t_{3(p_2)}$ and $v^t_{3(p_3)}$. Thus, the system of equations now reads

$$\begin{cases} 2.4 \cdot 10^9 \cdot \Delta t \frac{v^t_{3(p_3)} - 2v^t_{3(p_2)}}{2 + v^t_{3(p_3)}\Delta t} = 0 \rightsquigarrow v^t_{3(p_2)} = \frac{v^t_{3(p_3)}}{2} \quad \text{(equilibrium equation)} \\ \sqrt{\left(v^t_{3(p_2)}\right)^2 + \left(v^t_{3(p_3)}\right)^2} = 1 \quad \text{(arc-length condition)} \end{cases}$$

$$(K.32)$$

Replace $v^t_{3(p_2)}$ with $\frac{v^t_{3(p_3)}}{2}$ in the second equation of the above system, we obtain

$$v^t_{3(p_3)} = \pm\frac{2}{\sqrt{5}}, \quad \text{and} \quad v^t_{3(p_2)} = \pm\frac{1}{\sqrt{5}} \qquad (K.33)$$

Interpretation of results

The interpretation of the results is illustrated in Fig. K.3. The space of unknowns defined in the arc-length method (i.e. $x_{3(p_2)}$ and $x_{3(p_3)}$) is illustrated. The solution path in this one-dimensional example is a straight line. Point A is the given position at time t. For a given Δt, the position vectors at $t + \Delta t$ for the unknowns are

$$x^{t+\Delta t}_{3(p_2)} = x^t_{3(p_2)} + v^t_{3(p_2)}\Delta t = 1 \pm \frac{\Delta t}{\sqrt{5}}$$

$$x^{t+\Delta t}_{3(p_3)} = x^t_{3(p_3)} + v^t_{3(p_3)}\Delta t = 2 \pm \frac{2\Delta t}{\sqrt{5}}$$

The plus-minus symbol indicates that the arc-length method finds solutions at the intersection points (points B and C) of the solution path and the arc-length condition. Both points B and C are an arc length ($L = \Delta t$) away from point A. Point B is the preferred solution for the extension test simulation. Point C is another solution determined by the arc-length method that would be the solution for a compression test. It is obtained due to the second power in the arc-length condition (eq. K.31). This reveals that the arc-length method does not preserve the searching direction.

Due to the usage of arc-length condition, the arc-length L equals to the adopted Δt. That means, the coordinates $\left(x_{3(p_3)}^{t+\Delta t}, x_{3(p_2)}^{t+\Delta t} \right)$ in the solutions space is $L = \Delta t$ away from the coordinates $\left(x_{3(p_3)}^{t}, x_{3(p_2)}^{t} \right)$.

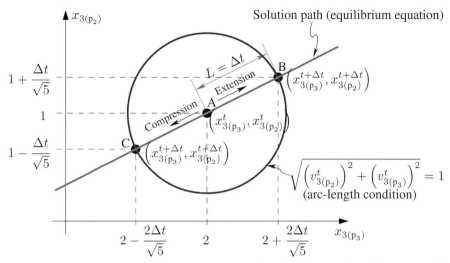

Figure K.3: Interpretation of the results obtained using the arc-length method. L is the arc-length.

K.3 Pseudo-arc-length method

In this section, it is to demonstrate how the pseudo-arc-length method is applied to solve this problem. For the details of the pseudo-arc-length method, see Section 2.7.4.4 (page 31).

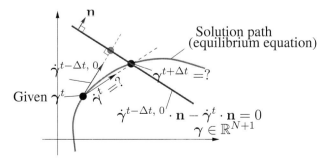

Figure K.4: Pseudo-arc-length method.

The pseudo-arc-length method differs from the arc-length method in:

1. A good guess $(\dot{\boldsymbol{\gamma}}^{t-\Delta t,\,0})$ is required for the tangent vector $\dot{\boldsymbol{\gamma}}^t$.

2. The used extra equation is replaced by an equation of a plane, which is $(\dot{\boldsymbol{\gamma}}^{t-\Delta t,\,0}\Delta t)$ away from $\boldsymbol{\gamma}^{t+\Delta t}$ and perpendicular to the good guess of $\dot{\boldsymbol{\gamma}}^t$, as shown in Fig. K.4.

The same parameterization is used in both methods. If we take the value of $\dot{\boldsymbol{\gamma}}$ obtained at $t-\Delta t$ (denoted as $\dot{\boldsymbol{\gamma}}^{t-\Delta t,\,0}$) as the guess for $\dot{\boldsymbol{\gamma}}^t$, the equation of the plane reads

$$\dot{\boldsymbol{\gamma}}^{t-\Delta t,\,0}\cdot\mathbf{n} - \dot{\boldsymbol{\gamma}}^t\cdot\mathbf{n} = 0$$
$$\rightsquigarrow \quad \boldsymbol{\nu}^{t-\Delta t,\,0}\cdot\mathbf{n} - \boldsymbol{\nu}^t\cdot\mathbf{n} = 0$$

with (eq. 2.84)

$$\boldsymbol{\nu}^t = \begin{bmatrix} v^t_{3(\mathrm{p}_2)} \\ v^t_{3(\mathrm{p}_3)} \end{bmatrix} \begin{array}{l} \leftarrow \mathbf{u}^t \\ \leftarrow v^t_e \end{array}$$

The derivation of the above equation is given in Appendix I. As a result, the system of equations for this problem is obtained (see eq. 2.92, page 32).

$$\begin{cases} \widetilde{y}_{3(\mathrm{p}_2)} = \nabla^{t+\Delta t}\cdot\mathbf{T}^{t+\Delta t}_{(\mathrm{p}_2)} = \widetilde{y}_{3(\mathrm{p}_2)}(v^t_{3(\mathrm{p}_2)},v^t_{3(\mathrm{p}_3)}) = 0 & \text{(equilibrium equation)} \\ \boldsymbol{\nu}^{t-\Delta t,\,0}\cdot\mathbf{n} - \boldsymbol{\nu}^t\cdot\mathbf{n} = 0 & \text{(pseudo-arc-length condition)} \end{cases}$$

K.3.1   Solution using the pseudo-arc-length method

Now we apply the pseudo-arc-length method to this example. The illustration of the pseudo-arc-length method applied to this problem is shown in Fig. K.5. Let the tangent vector

$$\boldsymbol{\nu}^t = \begin{bmatrix} v^t_{3(\mathrm{p}_2)} \\ v^t_{3(\mathrm{p}_3)} \end{bmatrix}$$

We choose the following (for example) as a guess to the tangent vector:

$$\boldsymbol{\nu}^{t-\Delta t,\,0} = \begin{bmatrix} v^t_{3(\mathrm{p}_2)} \\ v^t_{3(\mathrm{p}_3)} \end{bmatrix}_{\text{guess}} = \begin{bmatrix} 1 \\ 1.5 \end{bmatrix} \text{m/s}$$

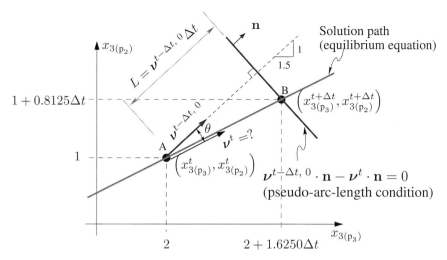

Figure K.5: Application of pseudo-arc-length method to solve this problem.

Given the guess of the tangent vector ($\boldsymbol{\nu}^{t-\Delta t,\,0}$), we have

$$\mathbf{n} = \frac{\boldsymbol{\nu}^{t-\Delta t,\,0}}{|\boldsymbol{\nu}^{t-\Delta t,\,0}|} = \begin{bmatrix} 0.5547 \\ 0.8321 \end{bmatrix}$$

The equation of the plane is then obtained:

$$\boldsymbol{\nu}^{t-\Delta t,\,0} \cdot \mathbf{n} - \boldsymbol{\nu}^t \cdot \mathbf{n} = 0$$

$$\rightsquigarrow \quad 1.8028 - \left(0.5547 v^t_{3(\text{p}_2)} + 0.8321 v^t_{3(\text{p}_3)} \right) = 0$$

As a result, the system of equation using pseudo-arc-length method reads (see eq. (K.32) for comparison with the arc-length method)

$$\begin{cases} v^t_{3(\text{p}_2)} = \dfrac{v^t_{3(\text{p}_3)}}{2} & \text{(equilibrium equation)} \\ 0.5547 v^t_{3(\text{p}_2)} + 0.8321 v^t_{3(\text{p}_3)} - 1.8028 & \text{(pseudo-arc-length condition)} \end{cases}$$

Solve the system of equations, we obtain

$$v^t_{3(\text{p}_3)} = 1.6250$$
$$v^t_{3(\text{p}_2)} = 0.8125$$

Interpretation of results

The solution procedure is illustrated in Fig. K.5. Given point A and Δt, the solution locates at the intersection (point B) of the solution path and the plane constructed using a good guess ($\nu^{t-\Delta t,\,0}$) of the tangent vector ν^t. For a given Δt, the position vectors at $t + \Delta t$ for the unknowns are

$$x_{3(p_2)}^{t+\Delta t} = x_{3(p_2)}^t + v_{3(p_2)}^t \Delta t = 1 + 0.8125\Delta t$$
$$x_{3(p_3)}^{t+\Delta t} = x_{3(p_3)}^t + v_{3(p_3)}^t \Delta t = 2 + 1.6250\Delta t$$

which is the intersection of the constructed plane and the solution path (i.e. point B).

Appendix L

Velocity of a Point on a Tilting Wall

A retaining wall is hinged on the bottom and tilts cyclically, as shown in Fig. L.1. Point A on the tilting wall is considered. The coordinates of point A is shifted with respect to the coordinates of the hinge ($\mathbf{x}_H$):

$$\bar{\mathbf{x}}_A = \mathbf{x}_A - \mathbf{x}_H \qquad (\text{L.1})$$

where $\mathbf{x}_A$ is the position of point A.

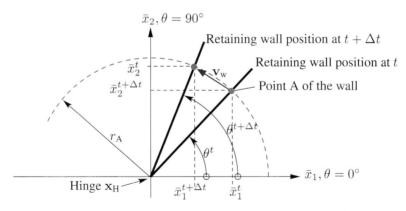

Figure L.1: Illustration of wall positions from t to $t + \Delta t$

The tilt angle on the retaining wall is defined in Fig. L.1. The retaining wall starts at $\theta = 90°$ and tilts at a angular speed $\dot{\theta}$ (scalar). For a positive value of $\dot{\theta}$, the wall rotates counter-clockwise about the hinge. Given θ^t and time increment Δt, we have

$$\theta^{t+\Delta t} = \theta^t + \dot{\theta}\Delta t \qquad (\text{L.2})$$

181

The velocity vector $\mathbf{v}_w$ of point A from time t to $t + \Delta t$ on the retaining wall is calculated as follows:

$$\mathbf{v}_w = \frac{\bar{\mathbf{x}}_A^{t+\Delta t} - \bar{\mathbf{x}}_A^t}{\Delta t} \tag{L.3}$$

with $\bar{\mathbf{x}}_A^t$ and $\bar{\mathbf{x}}_A^{t+\Delta t}$ being calculated as:

$$\bar{\mathbf{x}}_A^t = \left[\bar{x}_1^t, \quad \bar{x}_2^t \right]_A \qquad\qquad \bar{\mathbf{x}}_A^{t+\Delta t} = \left[\bar{x}_1^{t+\Delta t}, \quad \bar{x}_2^{t+\Delta t} \right]_A$$

$$\text{with} \begin{cases} \bar{x}_1^t = r_A \cos(\theta^t) \\ \bar{x}_2^t = r_A \sin(\theta^t) \end{cases} \qquad\qquad \text{with} \begin{cases} \bar{x}_1^{t+\Delta t} = r_A \cos(\theta^{t+\Delta t}) \\ \bar{x}_2^{t+\Delta t} = r_A \sin(\theta^{t+\Delta t}) \end{cases}$$

in which r_A is the distance from point A to the hinge:

$$r_A = |\mathbf{x}_A^t - \mathbf{x}_H| \tag{L.4}$$

Consequently, given $\mathbf{x}_H$, $\mathbf{x}_A^t$, θ^t, and $\dot{\theta}$, the velocity of point A on the tilting wall from time t to $t + \Delta t$ is computed as

$$\mathbf{v}_w = \frac{r_A}{\Delta t} \left[\cos(\theta^t + \dot{\theta}\Delta t) - \cos\theta^t, \quad \sin(\theta^t + \dot{\theta}\Delta t) - \sin\theta^t \right] \tag{L.5}$$

Appendix M

List of Symbols

Symbols that are often used in the main text are listed in the following. The page number is referred to the first time a parameter is defined or where it is well explained. The symbols used in Appendices are well defined therein, but might not be consistent with this list.

Symbol	Meaning
$\mathcal{B}$	barodesy for sand (constitutive model) (eq. 2.36, page 14)
$\mathbf{D}$	stretching tensor (eq. 2.5, page 6)
d^{of}	degrees of freedom (page 17)
$\mathcal{D}^{\mathrm{of}}$	degrees of freedom matrix (page 17)
e	void ratio (page 5)
Ξ	matrix of residuals of equilibrium equations at all particles in a system (eq. 2.43, page 16)
$\mathcal{E}$	vector of residuals of the equilibrium equations at a particle (eq. 2.42, page 16)
$\mathbb{E}$	matrix consisting of void ratios of all particles (eq. 2.44, page 17)
$\mathbf{g}$	- in mechanics: gravitational acceleration (eq. 2.1, page 5) - in the Levenberg-Marquardt's method, scaled residuals (eq. 2.69, page 24)
γ	a vector storing only unknown position components in arc-length methods (eq. 2.83, page 28)
$\dot{\gamma}$	tangent vector in arc-length methods (eq. 2.84, page 29)
h	step size used in substeps (eq. 2.98, page 34)
i_{e}	index for the numeration of the residuals of the system of equations $\mathbf{y} = y_{i_{\mathrm{e}}}$ (eq. 2.57, page 21)
i_{u}	index for the numeration of the unknowns in a system $\mathbf{u} = u_{i_{\mathrm{u}}}$ (page 21)
i_{w}	index for the numeration of the particles on the tilting wall (eq. L.5, page 182)
k	- in Newton's method: iteration counter (eq. 2.58, page 21) - in the nearest-neighbor search method: number of nearest neighbors (Section 2.3, page 9)
$\mathbf{L}$	velocity gradient (Section 2.4.2, page 12)

Continued...

Symbol	**Meaning**
$\mathbf{m}$	mapping vector that maps a matrix to a vector (Section 2.6.3, page 17)
$\mathbf{n}$	unit vector normal to a surface (eq. 2.122, page 39)
$\mathbf{n}_f$	unit vector normal to a free surface (eq. 6.1, page 85)
$\mathbf{n}_w$	unit vector normal to the retaining wall (eq. 6.8, page 88)
n_p	total number of particles in a system (page 5)
n_n	total number of neighbors at a particle (page 10)
n_w	total number of particles on the tilting wall (eq. L.5, page 182)
ν	vector storing only unknown velocity components, used in arc-length methods (eq. 2.84, page 29)
$\mathbb{R}$	matrix consisting of densities at all particles (eq. 2.44, page 17)
ρ	density (eq. 2.1, page 5)
$\mathbf{t}$	stress vector on a surface, traction (eq. 2.8, page 39)
Δt	time increment (eq. 2.9, page 6)
$\mathbf{T}$	effective Cauchy stress tensor (eq. 2.1, page 5)
$\dot{\mathbf{T}}$	effective rate of Cauchy stress tensor (eq. 2.31, page 14)
$\overset{\circ}{\mathbf{T}}$	objective rate of the effective Cauchy stress tensor (eq. 2.8, page 6)
$\mathbb{T}$	matrix consisting of stress tensors of all particles (eq. 2.44, page 17)
$\mathcal{T}$	matrix consisting of stress components of $\mathbf{T}$ of neighbors (eq. 2.28, page 13)
$\mathbf{u}$	vector consisting of only unknown velocity components in a system (eq. 2.50, page 18)
$\mathbf{v}$	velocity vector at a particle (page 12)
$\mathcal{V}$	matrix consisting of velocity vectors of neighbors (eq. 2.25, page 12)
$\mathbb{V}$	matrix consisting of velocity vectors of all particles in a system (eq. 2.44, page 17)
$\mathbf{W}$	spin tensor (eq. 2.6, page 6)
$\mathbf{w}$	vector consisting of only unknown positions in a system ($\dot{\mathbf{w}} = \mathbf{u}$) (eq. 2.52, page 19)
$\mathbf{x}$	position vector at a particle (page 12)
$\Delta\mathbf{u}$	Correction of the variables (iterates) in the Newton's method (eq. 2.61, page 21)
$\mathcal{X}$	matrix consisting of position vectors of neighbors (eq. 2.19, page 10)
$\mathbb{X}$	matrix consisting of position vectors of all particles (eq. 2.44, page 17)

Continued...

Symbol	Meaning
y	residuals of the left-hand side of the system of equations (eq. 2.51, page 19)
$\widetilde{y}$	residuals of the left-hand side of the system of equations (eq. 2.87, page 30)
$\mathcal{y}$	residuals of the system of equations parameterized using positions (eq. 2.53, page 19)
$\widetilde{\mathcal{y}}$	residuals of the system of equations parameterized using positions (eq. 2.86, page 29)